God & F

GOD & FORMS in PLATO

Richard D. Mohr

PARMENIDES
PUBLISHING

Originally published as *The Platonic Cosmology*
in 1985 by E. J. Brill, Leiden

This revised and expanded edition, with a new preface,
four additional essays, a new extension,
and updated bibliography, index locorum and author index
published in 2005 by Parmenides Publishing
in the United States of America

ISBN-13: 978-1-930972-01-8
ISBN-10: 1-930972-01-6

Printed in the United States of America

Library of Congress Cataloging-in-Publication Data

Mohr, Richard D.
God & forms in Plato: the Platonic cosmology / by Richard D. Mohr.—Rev. and expanded ed.
p. cm.
Rev. ed. of: The Platonic cosmology. 1985.
Includes bibliographical references and indexes.
ISBN-13: 978-1-930972-01-8 (pbk. : alk. paper)
ISBN-10: 1-930972-01-6 (pbk. : alk. paper)
1. Plato. 2. Cosmology, Ancient. I. Title: God and forms in Plato.
II. Mohr, Richard D. Platonic cosmology. III. Title.
B398.C67M64 2005
113.092—dc22 2005032661

Four essays by Richard D. Mohr reprinted with permission of the editors:

1. "Family Resemblance, Platonism, Universals" in *Canadian Journal of Philosophy*, Vol. VII, No. 3 (Sept 1977), pp. 593–600.
2. "The Formation of the Cosmos in the *Statesman* Myth" in *Phoenix* (Journal of the Classical Association of Canada), Vol. 32, No. 3 (1978), pp. 250–52
3. "The Divided Line and the Doctrine of Recollection in Plato" in *Apeiron* (a journal for ancient philosophy and science, by permission of Academic Printing and Publishing), Vol. XVIII, No. 1 (1984), pp. 34–41
4. "The Number Theory in Plato's *Republic* VII and *Philebus*" in *Isis*, Vol. 72, No. 264 (Dec 1981), pp. 620–27. Published by the University of Chicago Press.

1-888-PARMENIDES
www.parmenides.com

for

Robert W. Switzer

ᾧ συνδιαπονεῖν μετ' ἐμοῦ
τὰ πολλὰ οὐχ ἄηθες

"For whom it is not unusual to work through virtually everything together with me."

—Plato's *Sophist* 218b

CONTENTS

PREFACE TO THE REVISED AND EXPANDED EDITION (2005)

This book is an expanded edition of my 1985 book *The Platonic Cosmology* (Brill). Since the original work had to be re-keyed, I have taken the opportunity to update and revise its original chapters—correcting errors, discussing some of the secondary literature that has piled up in the intervening twenty years, and, here and there, trying to clarify positions and strengthen arguments. The positions taken and the conclusions reached, though, are for the most part unchanged. The original chapters have been supplemented with four additional essays that I wrote during the general period in which the book was written and that bear on some of its more general claims about Plato. These supplemental essays cover the nature of universals, a debate about temporal creation in Plato, recollection's relation to the Divided Line, and the nature of numbers in Plato. In addition, I have written for this edition a new essay, "Extensions," which draws together and develops some general thoughts that in the original book appeared only as wisps or stems. It tries to place Plato's cosmological commitments in the *Timaeus, Statesman,* and *Philebus* into a wider metaphysical context.

This is a book about how, for Plato, God makes the world. Unlike in the Catholic story about how God makes the world, God in this story does not make the world out of nothing. Indeed he doesn't make the world *out of* anything—even though he is chiefly characterized as a craftsman (δημιουργός) or, more simply, maker (ποιητής).[1] There is making and there is making. Plato's God has not read the second book of Aristotle's *Physics* and so does not know that one makes things by making them out of matter. On that view, one first finds some indeterminate stuff and then one imposes upon it a form, property, or shape where there was no form, property or shape before. Plato's God does not work like a carver who whittles an amorphous chunk of driftwood into a cube, nor like an artisan who pours molten brass into a mold to make a bell,

[1] Craftsman—*Timaeus* 28a6, 29a3, 41a7, 42e8, 68e2, 69c3, cf. *Philebus* 27b1. Maker—*Timaeus* 28c3, cf. *Philebus* 27a5.

ball-bearing, or "Bird in Flight." Neither does Plato's craftsman God make the world out of discrete components by assemblage. He does not make things by taking bits which have determinate characters (forms, properties, shapes), but no order among them, and by putting them into an arrangement create order. He does not work like a tile setter making a mosaic mural out of tessera, nor like a child on Christmas morning making a Ferris Wheel out of an Erector® Set—"some assembly required."

Plato's craftsman God or Demiurge, as he sets out to create the heavens, is not confronted with either formless matter or determinate bits. Rather he has been reading the neglected Milesian philosophers, Anaximander and Anaximenes, and knows that the phenomenal realm which confronts him is a booming, buzzing confusion of instances of determinable properties, say, of temperature, speed, thickness, height, and compactness, which—the instances, that is—career along scales of increase and decrease, slide along gradients of more and less, become hotter and colder, faster and slower, stouter or svelter, higher and lower, denser and rarer, drier and wetter, louder and softer, and the like. He improves upon the confusion by introducing measures into it, by eliminating from it excesses and deficiencies as the instances slide along the various gradients. This sort of making involves three sorts of skill and expertise. First, the craftsman himself has to have the ability to shift the instances of properties along the scales upon which they, on their own, slide. He needs to be like a doctor who knows how to raise and lower the temperature of an alternatingly chilled and fevered patient. Second, he must be able to fix the instance at a certain degree on the scale, like the doctor who arrests and holds the patient's temperature at a chosen degree. Finally, beyond these preliminary skills, the craftsman God needs to know what point on the scale of degrees is the right degree at which to fix the instance of the property. The good doctor needs to know that 98.6° F is the right degree at which to fix the patient's temperature. When he fixes the instance at the degree on the scale dictated by the standard, he has done his job. The craftsman God, like the accomplished doctor, needs a standard or measure. Yet this standard is not something that can be discovered simply by examining the scale of degrees upon which an instance of a property slides. The standard is off the scale. It is something like a meter stick, a thermometer, or a template, something by reference to which we assess and identify other things. And so if the Demiurge is to improve the world, he needs to have access to objects that serve as standards and which are not part of the phenomenal realm. These standards and measures Plato calls Forms or Ideas.[2] They provide content for

[2] Aristotle correctly detects that this need is one of Plato's motivations for hypothesizing Forms. He telegraphically calls it "the arguments from the sciences" (*Metaphysics* 990b12, 1079a8). See Mohr (1). Plato sees the need to make judg-

God's good intentions to craft. His good intentions and the world the way it is given to him are not sufficient for its improvement. Forms are essential to the nature of his craft.

And Forms are essential for there being any phenomena in the first place. Before the creation of the measured world, the phenomena are fleeting non-substantial images, like shadows on walls, images on water, and reflections in mirrors. To exist such images must have two things. They must have a medium, on which they appear, but which does not enter into them. Plato calls this medium the receptacle of space. And they must have originals which determine both what the images are images of and which must persist if the images are to exist. Plato calls these originals Forms. Without Forms there are no phenomena. Both God and Forms have distinct roles in Plato's metaphysics, roles that can not be reduced to elements of other philosophers' metaphysics nor be explained away as metaphysical luxuries within his own system.

When *The Platonic Cosmology* appeared in 1985, scholarly study of the *Timaeus*, Plato's chief cosmological work, was almost a wasteland.[3] In the mid-1950s, the dialogue with its ontological refulgence had been 'exiled' from the late dialogues to Plato's middle period where its expellers thought it could live in comfortable retirement along with the metaphysical excesses of "Socrates' Autobiography" in the *Phaedo*, the "Ladder of Love" in the *Symposium*, the analogies of Sun, Line, and Cave in *Republic* VI–VII, and the "Great Speech" of the *Phaedrus*.[4] Placed there, the dialogue could be forgotten without loss and the late dialogues would thereby be made safe for an agenda laid out initially in 1939 by Gilbert Ryle to the effect that in the late dialogues Plato had given up metaphysics for conceptual analysis—and praise be that he had done so.[5] For a couple of decades, the *Timaeus* attracted little critical attention beyond that which addressed issues immediately spawned by the dating controversy itself.

In 1985, the dating of the dialogue was still hotly debated.[6] But by the mid-1990s, the dialogue was firmly back in the late group.[7] Indeed a crit-

---

ments in general in the way we, with the aid of standards and measures, make precise judgments of lengths and weights. He does so as early as the *Euthyphro* (7b–d) and as late as the *Philebus* (61d–62b).

[3] So, the editor of *Phronesis* in a short notice of the book's appearance. *Phronesis* 31 (1986) 97.

[4] Owen (1). *Phaedo* 95a–102a, *Symposium* 210a–212b, *Republic* 505a–520c, *Phaedrus* 243e–257b.

[5] Ryle.

[6] Prior 168–93.

[7] Brandwood (1) 250, (2) 114. See Broadie (2) 187n7.

ical consensus now agrees with Brandwood's dating of it after the *Parmenides* and *Theaetetus*, dialogues in which the expellers thought Plato had rejected the theory of Ideas and outgrown metaphysics more generally, and immediately before the *Sophist*. This pressing of the metaphysical robustness of the *Timaeus* up against the supposedly lean, mean logic of the *Sophist* not surprisingly generated a great deal of cognitive dissonance among critics, most of whom, in the Anglo-American tradition at least, still harbored the philosophical sympathies of those who had exiled the *Timaeus* in the first place: Platonism is not about existence, Platonism is about predication—Plato is *Plato discursus*. For the most part among Anglo-American critics, the cognitive dissonance has been reduced *not* by interpreting other late dialogues to bring them into alignment with the *Timaeus* and its elaboration of the sharply dualistic theory of Forms, but by variously modulating and manipulating the *Timaeus*. This strategy has taken two general forms—dismissal and revision.

Dismissal. Those nominalist-inclined critics who believe that the main project of the late Platonic dialogues is careful conceptual analysis devoid of any metaphysical commitments, and who hold that Plato believes "that the activity of philosophy is more important than the specific views one holds"[8]—these critics have to remove the *Timaeus* from the forum of serious critical discussion. The dialogue's dualistic metaphysical commitments are only the half of it. Its dogmatic assertions, monolithic form, and *a priori* deductive arguments all seem diametrically opposed to the Socratic "spirit of open-ended exploration."[9] For these critics, the dialogue has to be massaged in such a way that it can be read as containing no serious doctrines: The "*Timaeus* offers the reader a rhetorical display, not a philosophical dialogue."[10] On this account, "what *Timaeus* says about 'being' and 'becoming', the Forms and 'reflections', the 'demiurge' and the 'receptacle', and the arguments he offers on these subjects" *can* simply be read as "rhetorical embellishment intended to impress Socrates and his other listeners."[11] This view fails to explain why Socrates would be interested in a mere rhetorical display, especially one that turns on ideas that Socrates had learned were false in the *Parmenides*. It also fails to explain why Plato himself would go to the bother of elaborating theories that were based on principles that he knew from the *Parmenides* to be false.

The position that the dialogue is just a myth gets such force as it has from the claim *within* Timaeus' discourse itself that Timaeus is spinning

[8] Nehamas (2) xlvi; similarly McCabe (1) 18–21. And see, in general, the essays in Gill and McCabe, and in Press.

[9] Cooper xxii.

[10] Cooper 1224.

[11] Cooper 1225.

a likely myth or likely story (29b–c).[12] But the status of myth cannot be extended to Timaeus' whole discourse, for what Timaeus says more fully here is that discourses will have their degree of certainty conferred upon them by their objects. Discourses will necessarily be only likely to the extent that they are about the likenesses that make up the phenomenal world, but discourses about the Forms, since the Forms are abiding and stable and accessed through reason, can be incontrovertible and irrefutable. And indeed Timaeus gives us an example of just a such claim, one presumably based on the content of the Forms, but which in any case he thinks has the status of being incontrovertibly true and, surprise, it is a claim about the very metaphysical system of the *Timaeus* itself, at a minimum a claim about the relation of Forms to space (52c7–d1).[13] So if we take seriously Timaeus' claims about the status of his account of the phenomenal world, we must also take seriously that he intends his various claims about the Forms and their relations to the rest of the furniture of the universe to be taken as accurate and certain.

A variant of the strategy that dismisses the *Timaeus'* dualism is a strategy that evades it. Critics will simply disregard or bypass what they find philosophically annoying in the dialogue, act, for instance, as though the theory of Forms is not part of the dialogue. One critic says that "the reader must take the *Timaeus* with several pinches of salt" and then uses this interpretive principle as a license to treat what is found interesting about the dialogue as representing Plato's actual thoughts, but to overlook what is found inconvenient or hopelessly old-fashioned.[14] More generally, *Timaeus* scholarship of late tends to focus on topics that don't require taking a stand on some of the larger or weightier or loaded metaphysical issues in the dialogue. So, for example, we now find lots of concern with methodological issues in the dialogue.[15] And a whole cottage industry has grown up around the prospect in the dialogue of people becoming like God, though little usually gets said there about what the god is like.[16] The motto for this interpretive trend in *Timaeus* scholarship could read, "Anything but Forms, please."

Revision. Those critics who dismiss or evade the metaphysical content of the *Timaeus* tend to be nominalists. In contrast, those critics who give

---

[12] Cooper 1225.

[13] The text: "As long as one thing is different from another, neither of them ever comes to be in the other in such a way that they at the same time become one and the same, and also two" (Timaeus 52c7–d1). At a minimum, the sentence holds that Forms cannot be in the receptacle of space and the receptacle cannot be in the Forms, since a Form and the receptacle are each on its own fundamentally one in number.

[14] McCabe 162.

[15] For example, Strange, Osborn.

[16] Sedley 798–806, Annas (4) chapter 3.

the *Timaeus* revisionist interpretations tend to be Aristotelians. Here the strategy is to grind down the dialogue's grand metaphysical commitments to tamer, Aristotelian ones, and when that reduction can't quite be entirely accomplished, to leave the metaphysical detritus without a philosophical function. The most recent monograph on the *Timaeus* is honest about this approach: "From the point of view of articulating Plato's teleology I have thought it instructive to place the *Timaeus-Critias* within the context of Aristotle's philosophy."[17] Now, if one looks through a lens with a tree painted on it, one will see a tree; if one looks through a pink-colored lens, one will see pink; if one looks through an Aristotelian lens, one will see Aristotle.[18] So for example, the Demiurge on this reading has to be explained away as a distinctive metaphysical entity since Aristotle's teleology, whether of *Physics* II or *Metaphysics* XII, is not explained by appeals to efficient causes, like particular makers, builders, or craftsmen. So instead of having a distinct philosophical function and status, the Demiurge is sublated into an abstraction. He is just a symbol of craftsmanship in general.[19] And his craftsmanship is of "order and hence goodness."[20] But one can produce garden-variety order, mere assemblage, without appeals to Forms or models: take pencil and paper; draw a square; then divide it into four squares; then divide the lower right square into four squares, then divide the lower right square of that four-square into four squares, and so on. If this sort of ordering is all the Demiurge does, he can be kept busy producing goodness for a very long time—ever increasing goodness by ever increasing orderliness. No. Forms are needed for demiurgic crafting in order to put a limit on excess, and when that limit is achieved, when exact correspondence to a model is reached and fixed, crafting stops.

Or take the *Timaeus'* major new addition to the ontology of the middle period Platonic dialogues—the receptacle of space. In the last few decades, critics have overwhelmingly agreed with Aristotle that Plato's space is a forerunner of Aristotle's own principle of matter, that it serves as and is philosophically motivated as being a principle of individuation and substrate for change.[21] But if space is a material component of the phenomena that stabilizes and locates them so that their natures may shine forth on their own, then there is not much that the Forms can do

---

[17] Johansen 5.

[18] The recent trend to grind Plato down into an Aristotelian is not limited to *Timaeus* scholarship. Gail Fine has been doing it for the Plato of the *Republic*: "I find it illuminating to look at Plato's metaphysics through Aristotelian lenses." Fine (2) 41. Look through a lens with a tree painted on it, ...

[19] Johansen 86.

[20] Johansen 72.

[21] Kung, Algra, Zeyl, Miller, Johansen.

for the phenomena. On this reading, the phenomena in the receptacle of space have become Aristotelian substances, composites of form and matter, and Forms themselves are left as so much excess ontological baggage. No. Forms are needed because the phenomena have only the status of non-substantial images fleeting across a medium which does not enter into them, and so Forms are needed so that the phenomena may both exist and be what they are as reflections of Forms.

It is my hope that the re-issuing of my views on these controversies may act at least a bit as a corrective to some of the trends in recent Platonic scholarship mapped here. Plato is far more interesting than what we would be led to believe from most of the recent literature.[22] We live in a period when, to make Greek philosophy respectable by contemporary philosophical standards, it has been made boring. The borification of Greek philosophy extends beyond the Aristotelianizing and nominalizing critics of Plato. We see it in the banishing of the universal flux doctrine from Heraclitus. We see it in the banishing of prime matter from the *Physics*. We see it in the erasure of non-substantive individuals from the *Categories*. All these moves are made by well-intended critics trying to make Greek philosophy street-legal.

I'm pleased to be part of the Parmenides Publishing project, which seems committed to making Greek philosophy an adventure again. It is nearly miraculous that my Plato book should be having a chance at a second hearing. Its first was a little strange. The book was published in the summer of 1985; reviews began appearing in 1987, but by the summer of 1989, it had sold out.[23] By the arrival of that news, though, I was already off on another adventure, trying to save the world with a series of books on lesbian and gay social and legal policy (*Gays/Justice*, 1988; *Gay Ideas*, 1992; *A More Perfect Union*, 1994; *The Long Arc of Justice*, 2005) and, along the way, to save myself, I began writing as well on the arts, decorative arts, and architecture (*Pottery, Politics, Art*, 2003). These additional lives did not leave much time for writing on Plato, though I continued teaching advanced undergraduate courses and graduate seminars on the Ideal theory and built up a cache of new or at least more developed thoughts on Plato, some of which have found their way into this edition's new chapter "Extensions." I regret that I've not been able to deal as thor-

[22] There *are* some books that run counter to the general trends. I applaud Patterson and Prior, whose Plato books appeared the same year as *The Platonic Cosmology*, and Silverman, a rare twenty-first century hold out for dualistic unitarianism. Those who view Plato as valuable and coherent generally take seriously the metaphors of light and image and of Forms as models or standards from the central books of the *Republic* and the *Timaeus*.

[23] Letter from J.G.Deahl, Editor, Brill, to Richard Mohr, July 31, 1989.

oughly with the secondary literature that has appeared since 1985 as the original edition did with that which appeared before.

I'd like to thank those scholars who took time to write reviews of the earlier edition: Elizabeth Belfiore, William Prior, Elizabeth Asmis, John Dillon and E. N. Lee.[24] I can repay the favor at least a little now by having taken their criticisms into account in the revisions. The greatest influences on my writings on Plato have been Harold Cherniss, R. E. Allen, and E. N. Lee. The book, especially with the supplemental chapters, hopes to be a contribution to the beleaguered American unity school of Platonic thought, which was launched in 1903 with Paul Shorey's *The Unity of Plato's Thought* and of which Cherniss, Allen, and Lee have been chief members.

Some of the book's chapters have roots going clear back to my doctoral thesis, *Studies in Plato's Cosmology*, which I completed in 1977 at the University of Toronto, with T. M. Robinson as supervisor and R. E. Allen as advisor. I thank Robinson especially for setting an approach to texts that I have followed and for having made possible an academic career for me.

For help with Greek matters in the expanded edition, I am much indebted to David Sansone and Zina Giannopoulou. Many thanks.

I would also like to thank the partners with whom I have lived during the writing of the component parts of the project, Susan J. Slottow and Robert W. Switzer. This book is dedicated, as have been all my others, to my partner and now, thanks to Canadian legal reform, husband, Robert W. Switzer.

---

[24] Elizabeth Belfiore, *Religious Studies Review* 31 (1987); William Prior, *Journal of the History of Philosophy* 25 (1987); Elizabeth Asmis, *Phoenix* 41 (1987); John Dillon, *Canadian Philosophical Reviews* 8 (1988); E. N. Lee, *Ancient Philosophy* 11 (1991). *Ancient Philosophy* had commissioned a review from Joan Kung, but unfortunately she died before finishing it.

FROM THE ORIGINAL PREFACE (1985)

The book may be read in several ways. First, the arrangement of chapters is by the chief text that is discussed in each. Each is self-contained and may be read independently of the others. Those whose interest lies in the meaning of specific passages are referred to the table of contents and the *index locorum.* A review of the *status quaestionis* is provided for most passages discussed.

Second, the chapters dovetail in such a way to cover collectively most of Plato's cosmological writings, and so the collection may be read straight through as something approaching an interpretative commentary to the allied texts which make up Plato's cosmological corpus.

Third, the book may be read selectively for specific Platonic doctrines. The chapters discuss texts not merely to the end of explicating difficult and much debated passages but primarily in order to solve problems in the Platonic cosmology or to clarify difficult themes or doctrines which extend beyond localized texts. The reader is therefore invited to look at the introductory "Themes and Theses" section, which collects together the various results gleaned from particular analyses and is offered as a sort of reader's guide to the Platonic cosmology taken as a body of doctrines or philosophical perspectives. The section states the nature and range of problems discussed in the chapters and lays out the systematic relations between the chapters. It sketches in broad outline what I take to be Plato's major cosmological commitments, the details and exposition of which make up the body of the book.

INTRODUCTION: THEMES AND THESES (1985)

Spurred by the unanticipated discovery in 1964 of a uniform background radiation throughout the universe—a ghostly vestige of the explosion with which our universe began—cosmology has in the ensuing years risen from near total neglect as a discipline to become a leading discipline within the physical sciences. It has also captured the popular imagination as the result of a certain curiosity that humans have about origins. Scientists, however, when forthright, admit that current cosmology will not be responsive to the human hankerings and sense of wonder which fuel our curiosity about ultimate questions. For cosmology at present is a *special* discipline and an *empirical* one. Present cosmology has a specific subject: in what kinds did energy and matter exist and under what laws were they governed in the first instant of our universe when it existed in size smaller than a latter-day electron. And the methods of current cosmology are those of any science—though the laws of cosmology are gradually exhibiting such a high degree of simplicity and coherence, that a great deal of theoretical work can be done with an empirical base which seems so narrow when compared with the more familiar sciences that the impression is sometimes left that cosmology is mere speculation.

For Plato—for good or ill—cosmology is neither a special science, as it is already by the writing of Aristotle's *De Caelo,* nor is it empirically based. The phenomena for Plato offer problems rather than answers to the inquiring mind—they do not verify theories, rather theories "save" them. And the subject of cosmology is for Plato not limited to the sensible and material. It studies all things in all their facets, limited only as being seen from the perspective of the origin of the formed universe. Herein lies the particular importance of Plato's cosmology.

Though Plato's particular views about the start of things may be of little interest—maybe even to Plato himself, if one supposes that the creation story of the *Timaeus* is not intended to be read literally—Plato's cosmological writing, especially the *Timaeus,* is the one place in Plato's work where all of the "furniture" of Platonism—gods, souls, Ideas, space, properties, natural and artificial kinds—are seen related each to all within a single frame, and where all the major branches of speculative thought in Plato—epistemology, metaphysics, theology, physics, and to an extent logic and ethics—are seen, if not in all their detail, at least in their significant relations to each other. And so the study of Plato's cosmology casts light on all of Platonism.

Metaphysics and the Characteristics of Platonic Forms. The first two chapters in large part explore the consequences of viewing Platonic Forms as standards, measures, exemplars or paradigms—a view in the Platonic dialogues most prominent in the *Timaeus,* but not absent even from other late dialogues. I suggest that Platonic Forms are not just the fundamental individuals of Plato's ontology—a position to which all would agree—but moreover, taken as standards, they are fundamentally individuals, rather than things qualified. Each Form has a core conceptual content which can not be analyzed into other Forms. The relations between Forms are minimal in importance compared with their peculiar content and the examination of these relations, even though they are necessary ones, does not produce essential definitions of Forms. Further, the names of Forms are not even to be construed as disguised definite descriptions but rather are more like Russellian logically proper names (distinguishable "this-es" and "that-es"). In consequence, the theory of Forms avoids a number of logical perils. In particular it is not subject to the vicious infinite regress of the Third Man Argument, since Forms fail to have the properties which they are used to define in other things and so further the question does not arise of how a Form and its instances are similarly disposed, a question which would require for its answer an appeal to a new Form in order to explain the similarity.

Knowledge, on this model of Forms, will turn out not to consist of conceptually locating a Form in the network of Forms in which it is embedded. Rather Platonic knowledge will be a kind of direct acquaintance of a Form in isolation, though the acquaintance will not be the kind of acquaintance one has on a representational realist model. Rather, the "seeing" of Forms is to be understood on a rough analogy with the way in which without a moment's reflection and really without doubt we recognize or "spot" an individual person as the individual he is.

The Platonic "to be" as applied to Forms turns out to be an existential sense ("to exist"), rather than a predicative sense ("to be F") or a veridical sense ("to be true" or "to be the case"); the existential sense though is not our modern sense "to be an instance of a concept" or "to be the value of a variable," but rather falls within the constellation of senses "to be actual," "to be there in such a way as to provide an object to point at," "to present itself." When a Form is said "to be really, completely or purely," the compounded designation will mean "is this way (actual, self-presenting) *on its own* or *in virtue of itself*" or "is there to be picked out independently of its relations to anything else" (chapter 1).

When Plato describes the Forms as "always existing," he means that Forms, in consequence of their being standards and nothing else, possess a type of timeless eternity, as opposed, that is, to possessing that cluster of properties: perpetual endurance, everlastingness, unchanging duration or sempiternity. The specific sense in which Forms are timelessly eternal is that they fall outside the category of things which are possible subjects for judgments of dates and durations (chapter 2).

Theology and the Platonic Demiurge. In the *Timaeus* Plato marks out as one of the central components of his cosmology a god who is unique and rational and who is described essentially as a demiurge, that is, a craftsman. This is the same Demiurge who appears in the *Republic* VII, *Sophist, Statesman* and *Philebus* and is a different one than the divine craftsman of the *Laws X* (see chapter 11). The chief project of the Demiurge of the *Timaeus* is directed to an epistemological end rather than a moral or aesthetic one. What the Demiurge does primarily is to introduce standards or measures into the phenomenal realm by imaging as best as he can in recalcitrant materials the nature of Forms where Forms are construed as standards or measures. This task requires the reproduction of two sorts of properties of Forms. First, the Demiurge must make a phenomenal particular conform as closely as possible to the Idea of which it is an instance by eliminating any degree to which the particular exceeds or falls short of exact correspondence to the Ideal-standard. Second, he must try to reproduce in the phenomena, if possible, the properties which are requisite to a standard's ability to serve as a standard—importantly among which are uniqueness and some sort of permanence. The Demiurge's project so conceived is the key to understanding the unique world argument and the discussion of eternity, passages around which revolves the rest of the dialogue's first half. The Demiurge makes the world unique and in some sense eternal because the world as a whole is in some way to serve as a standard (chapters 1 and 2).

The introduction of standards into the phenomenal realm has the consequence that if one has only the phenomenal realm (and not also the Ideal realm) as the object of one's cognition, nevertheless one is not limited to making judgments based simply on merely relative comparisons of greater and less, but can also make the sort of exact judgments that are only possible by making appeals to standards or measures. Indeed, the immanent standards serve as the objects of true opinion in the same way that Forms (as standards) serve as the objects of knowledge. The Demiurge, by introducing immanent standards into the world, improves it by making it more intelligible.

The Demiurge of the *Timaeus* is thus unlike most of the gods of the Western theological traditions. He does not primarily serve as a source of order understood as arrangement or assemblage, and so Plato's God here differs from Anaxagoras' Mind and its later variants—the gods of the teleological arguments for divine existence. Nor is the Demiurge here primarily viewed as a source of motion as are Empedocles' Love and Strife and the gods of the cosmological arguments for divine existence.

The God of the *Laws* X differs in several important respects from the God of the *Timaeus.* First, he is shown to exist as the result of the first formulation in Western thought of the cosmological argument (chapter 8). Second, his incursions into the world are primarily aesthetic and moral in intent (chapter 11). Further, his works are not impeded by a certain cussedness in his materials. His fully rational plans are not sometimes defeated by physical necessity, as are some plans of the Demiurge of the *Timaeus.* In consequence, Plato must advance in the *Laws* X—as he need not have done in the *Timaeus*—a theodicy, an explanation of the existence of evil in the world, given that its God is both good and untrammeled. Plato's solution to this problem, like his formulation of the cosmological argument, is taken over in variant forms into Christian thought: the world as a whole is in fact better if evil is allowed to exist in some of its parts than if there is no evil in it at all. For the possibility of actual evil in the world allows men to have free-will and a world with men of free-will is better than a world without them (chapter 11).

It is usually supposed that the rationality of the Platonic god(s) commits Plato to viewing God as having the status of an ensouled creature. I argue that Plato is not burdened by such a commitment and that Plato has good reasons to want his God to exist independently of soul. Insofar as Plato considers the Demiurge to be a necessary existent part of whose essence is rationality, then it behooves him not to have that rationality inhere in soul. For soul has a potential for irrationality, since it is subject to the buffetings and disruptive incursions of the bodily. The Demiurge's essential rationality is assured only if he is not a soul (chapter 10).

Bodies and Souls. One of the odder sounding commitments of the Platonic cosmology is to the existence of a soul for the whole world. Plato believes that quite apart from the individual human and animal souls which occupy bits of the world, a single rational soul is stretched throughout the whole of the physical realm. This World-Soul has two quite specific functions in the economy of Platonism. First, it serves as an immanent standard of the Demiurge's creation. Its uniform revolutions, as imaged in the revolutions of the heavens, serve as a model or standard for people's rational capacities. When the irregular circuits of a person's soul begin to imitate the regular circuits of the World-Soul, as the person becomes like *this* god, the person's soul comes to have the ability to make sense of the world around it, has the capacity to make true judgments both of the phenomena and of the Forms. This function of the World-Soul is emphasized in the *Timaeus* (chapter 1).

Second, the world soul has the function of maintaining the homeostatic conditions, the measures and proportions, of the Demiurgically created world against a natural tendency of the corporeal to be chaotic and disruptive. This function explains why Plato can without paradox claim both that everything is in flux and yet that measure and proportion are everywhere manifest in nature. The World-Soul operates rather like a governor on a steam engine: the governor regulates the motion of the machine in such a way that the machine's self-sustained and independently originated motions, which owing to unpredictable conditions of combustion tend to run off to excess, are nonetheless uniformly maintained and do not destroy the machine itself (chapter 9). This is the function of the World-Soul emphasized in the *Statesman* myth and *Philebus*, though it is consistent with the main function of the World-Soul in the *Timaeus*.

The phenomenal or corporeal realm, I argue, is on its own in constant motion. The explanation of the flux of the phenomena is completely mechanical and physicalistic, ultimately resting on a theory of weights in which each primary body (earth, air, fire, water) has a natural region towards which it tends to move. However, given an initial random distribution of primary bodies and given certain properties of their shapes (chapter 5), when the bodies begin to interact as the result of their motions toward their respective natural regions, their motions are always deflected or even reversed, and the result is not a separation of kinds but a perpetual flux and intermingling of bodies (chapter 6).

The theory of flux is advanced without appeal to a conception of souls as causal agents and this feature sets the *Timaeus* (chapter 6) and the *Statesman* (chapter 7) at odds with the argument for the immortality of the soul in the *Phaedrus* and the argument for God's existence in *Laws* X. In these arguments, it is claimed that self-moving souls are the causes of *all* motions. Many different attempts have been made to square these four dialogues on this issue. I argue that none succeeds (chapters 6–8 of the original book, plus the supplemental chapter 13).

Space and the Status of the Material World. For Plato the phenomenal or material realm has a double aspect. On the one hand, the phenomena are in flux; on the other, they are images of the Ideas. To the extent that the phenomena are in flux their very intelligibility and even existence are drawn into doubt. A thing in flux cannot be identified with respect to its kind; indeed nothing whatsoever may be said of it, either that it is "of such and such a type" or even that it is a "this" or "that." However, in their aspect as images of Ideas, the phenomena may at least be identified according to kind by being severally referred to the Ideas of which they are images. In order to save, in this way, the phenomena's intelligibility (however partial and derivative it may be), Plato hypothesizes the existence of space or the receptacle as a medium over which the phenomena as images may flicker (chapters 3 and 4).

In consequence of space serving essentially as a medium or field for receiving images, it differs in a number of important ways from Aristotle's matter, of which it is often mistakenly considered a forerunner. Plato's space, unlike Aristotle's matter, is not 1) a material cause of phenomenal objects or that out of which phenomenal objects are (made), 2) a principle of individuation for phenomenal objects, 3) an ultimate subject of predication for statements about the phenomenal realm, nor 4) a substrate for phenomenal change.

Space is however a principle of existence for the phenomena. As images, the phenomena lead a doubly contingent existence; they depend for their existence upon the persistence both of the originals of which they are images and of the medium in which they appear. They are, however, in this way saved from the threat of utter non-existence. And so Plato's conception of phenomena clinging to existence by appearing as images in a permanent medium resolves the problem, left over from the *Republic* of how becoming holds a middle ground between real being and utter non-being (chapter 3).

(2005)

For Plato, members of the disparate phenomena may be correctly called by the same name without strictly speaking sharing common properties. Rather each is what it is by a relation to its Form and this arrangement leaves open the possibility that phenomenal objects with the same name form family resemblances—without having a single defining thread that runs through all of them, the same one to the next, by which they are correctly called what they are called. The phenomena may stand to the Forms as musical variations stand to the theme on which they are variations. This possibility means that if we were magically able to slice the Forms off from Plato's ontology, we would not be left with an Aristotelian world of individuals that are composed of something that makes them each one in number and something else that makes them the type of thing they are and that they have in common with other things of the same type. Neither the Platonic "upstairs" nor the Platonic "downstairs" consists of Aristotelian objects (chapter 12).

For Plato, we are not able to make true judgments of the phenomenal world just by using the resources that the phenomenal world presents us. True opinions are not the result of contentless mental manipulations of content provided to us by sense perception. To make true judgments across the range of normative, numerical, and relational concepts, we need mental content to help pick out the instances of these concepts in the phenomenal world, since these instances are not immediately accessible to the senses. I suggest that these mental concepts are the objects of the third level of the *Republic*'s Divided Line and form a much needed link between the Divided Line and Plato's doctrine of recollection (chapter 14).

The third level of the Divided Line does not consist of "mathematical intermediaries," entities to which Aristotle thought Plato was committed in order that mathematical calculations and geometrical proofs would have objects of reference. On the one hand, according to Aristotle, such referents must be eternal and perfect like the Forms—but unlike the phenomena. On the other hand, mathematical objects must be plural like the phenomena—but unlike Forms, which are each one of a kind. There must be at least two twos if $2 + 2$ is to equal 4. So calculations and proofs cannot be about either Forms or particulars; they must be about some mysterious entities that are plural but perfect. Call them mathematical intermediaries—or so it goes on Aristotle's understanding of Plato. Finding mathematical intermediaries or their kin in Plato is an on-going, ever popular parlor game among Plato scholars. But Plato did not believe in such things. For him there are just two sorts of numbers. On the one hand, there is the Idea of number, which is the natural number line. Its numbers are defined by their position in a series, rather than by units that purportedly make them up. They don't have those. They are the ordinals and they are not subject to mathematical calculations—addition, multiplication, and the like. On the other hand, there are numberings of phenomenal objects, two objects, twelve objects, etc., all countable. The phenomena have cardinality, and they have this perfectly. For Plato, some instances of Forms are perfect particulars, and in the case of the Forms of integers, all their instances are perfect particulars. If mathematical calculations require objects, the perfectly countable phenomena will do. This dualistic and exhaustive understanding of numbers is advanced in both *Republic* VII and the *Philebus,* in each case as a model for understanding Forms in general. Plato's ontology did not shift from middle dialogues to late (chapter 15).

PART ONE

THE WORKS OF REASON

Timaeus 27d–47e

ONE

Divinity, Cognition, and Ontology: The Unique World Argument*

SYLLABUS

I. INTRODUCTION

At *Timaeus* 30c–31b, Plato gives an argument to show that the world is unique, that is, that it is one in number and the only instance of its kind.[1]

---

* This chapter was first published in *The Platonic Cosmology*.

[1] The text: "In likeness to which of the Living-things did he who fashioned it fashion it? We must not suppose that it was anything the nature of which has the status of a part.[a)] For no image of an incomplete thing is ever good. Rather let us posit that the world is similar, above all things, to that Living-thing of which all the other Living-things severally and in groupings are parts. For this Living-thing has, contained within it, all the intelligible Living-things, just as this formed world contains us and all other creatures which have been formed as visible things. For the god, wishing to make this world as similar as possible to that intelligible thing which is best and in every way complete, made it one visible thing, having in itself all the living things which are akin to it in virtue of their nature.[b),c)]

"Have we, then, correctly called the heaven one in number, or would it be more correct to call it many or even infinite in number? It is one, since it was crafted in accordance with its paradigm. For that which embraces whatever living things are intelligible could not be one of two. For in that case it would be

This chapter will give the argument a new interpretation, an interpretation on which the argument turns out to be intellectually respectable. Previous interpretations either are vague about just what the argument is or commit Plato to one or another logical fallacy, conceptual confusion, or metaphysical shipwreck.[2]

Further, I will suggest that the argument, when properly construed, offers a key to understanding major parts of Plato's metaphysics. It throws light on the workings of the Demiurge, Plato's singular rational craftsman God, who also appears in the *Republic* VII (530a) and at crucial junctures in *all* of the late so-called critical dialogues which expound positive doctrines, that is, the *Sophist* (265c–266d), *Statesman* (269c–273e) and *Philebus* (26e–27b, 28d–30e). The argument also throws light on the nature of the objects of cognition for Plato; it aids in understanding both the way in which Forms are the objects of knowledge and the way in which the phenomenal realm is the object of true opinion. Importantly it also throws light on the nature and status of the Platonic Forms, especially with respect to their nature as paradigms. The argument has a further interest in that it is one of a very few passages in the *Timaeus* where there is any integration between Plato's metaphysical speculations and the specific scientific enquiries which make up the bulk of the dialogue (e.g., microphysics, geophysics, astrophysics, mechanics, hydraulics, op-

---

necessary for there to be around these two in turn yet another Living-thing, of which these two would be parts. And thus it would be more correct to say that this world is likened, not to those two, but to the one that would embrace them. Therefore, in order that this world should be similar to the entirely complete Living-thing in respect to its singularity[d)]—for this reason, he who makes good worlds did not make two or an infinite number of them, but this heaven has come to be and is and ever will be one and unique" (30c3–31b3).

a) for εἶδος as a mere ontological place-holder, see, for example, 48e3. To translate μέρος here as "species" would be at best misleadingly specific and at worst question begging.

b) my guess is that the restrictive subordinate clause is simply periphrastic for "which are visible" (cf. 30d1).

c) the whole participial clause is merely descriptive rather than causal; indeed it is only incidentally descriptive, as can be seen from 39e3–5, where the introduction of particular kinds of animals into the world is viewed as a project quite independent of the project of making the world unique (see below n9).

d) The word is μόνωσις, which seems to be a Platonic coinage. This is the only occurrence of the word in Plato. See section IXBi for its sense.

[2] There are three currently live interpretations of the unique world argument, those of Keyt (1), Parry, and Patterson (1). Earlier commentators are simply not very clear on exactly how they suppose the argument is to be taken, even when they suppose that it makes sense. See, for example, Cornford (2) 40–43.

tics, metallurgy, botany, physiology, and pathology). We shall see that the argument both has an effect on the sorts of biological claims Plato makes and, in part, is influenced by his biological views.

Getting good results from the argument requires a fair amount of work. The argument is full of oddities, not the least of which is that the kind of which the world is the only instance is living-thing-in-general. The world as a whole is one living thing, the World-Animal, but it is not any particular species or specific kind of living thing. Further the argument is as obscure as it is brief. Broadly it runs: the Demiurge makes the world as like as possible to its model; the model is unique; and so, the Demiurge makes the world unique. Further it is argued that the relevant model here is the Idea of animal-in-general, the relevance of which is to follow from the nature of the relations which this Idea holds to other Ideas, specifically to the Ideas of other types of living things. These Ideas are said to be parts of it.

The argument is remarkable in that it presupposes, for an understanding of its premises, so much of the Ideal theory. In fact I suggest that the passage can be taken as a sort of case study in the Platonic metaphysics. For the argument presents the theory in its starkest and most technical vocabulary, *presuming* that the reader knows the meanings of the technical vocabulary from elsewhere. More specifically, the passage is remarkable in that it is one of only a few passages in the Platonic corpus which deal simultaneously with relations *both* between a Form and its phenomenal representations *and* between a Form and other Forms.[3] On the one hand, Forms are portrayed canonically as paradigms (31a4) and the relation of Form to particular is portrayed canonically as that of likeness (30c3, 7, d3, 31a8, b1) or more precisely as the relation of original to image (30c5). On the other hand, there is also to be found in the passage the technical vocabulary from the *Phaedrus* (264e–266b), *Sophist* (251a–259d), *Statesman* (262b–264b, 285a–b), and *Philebus* (16c–18e) with which Plato elaborates his views of the relations of more general to more specific Forms. It is important to note that when in the unique world argument Plato says that some Forms are parts of others (30c4, 6, 31a6) and that some Forms contain or encompass (περιλαβόν, περιέχον) other Forms (30c8, 31a4, 8), these expressions have technical senses in the Platonic metaphysics and Plato is assuming that the senses are known to his read-

[3] The other candidates for passages which state positive doctrines about both inter-Ideal relations and relations between Forms and their instances are the final argument from the *Phaedo* (especially 104b–105b), the discussion of dialectic in the *Phaedrus* (265c–266b, on the assumption that 265c9–d1 includes a backwards reference to 249b–c), and possibly the passage of the *Statesman* on non-sensible images (285c–286b), on which, though, see Owen (3).

ers so that he can refer to them in as brief a compass as he does to the relations of Forms to instances. The meanings of these phrases are not to be gleaned from or simply construed for the sake of this passage in isolation. I will also be suggesting that when Plato in the argument repeatedly speaks of Forms as "entirely complete" (30c5, d2, 31b1), he is also referring to a technical doctrine (sections V, VII).

This chapter makes no presumptions on the issue of whether the *Timaeus* was composed before or after the *Sophist-Statesman* group, in which the parts-of-Forms doctrine is most fully articulated. However, I will suggest that the unique world argument is unintelligible without a correct understanding of the parts-of-Forms doctrine as it is articulated in this group (section V). But whether the elliptical references to that doctrine in the argument are a foreshadow or an echo of that doctrine need not concern me, if I am correct in showing that the argument is indeed unintelligible without the doctrine. Those critics who, following Owen, suppose that Plato abandoned the transcendental theory of Forms in the late 'critical' dialogues will no doubt bridle at this way of proceeding.[4] However, if I can make my point, minimally these critics will have to admit that the parts-of-Forms doctrine and more generally the doctrine of the interweaving and blending of Forms, doctrines which they would like to see as superseding the transcendental doctrine, are at least compatible with that doctrine, as the presence of both sets of doctrines in the critically-neglected *Phaedrus* would independently seem to bear witness.[5]

It would not be extreme to suggest that in the unique world argument Plato's deep intent is indeed to show off the machinery of the Ideal theory. He seems to be using the passage as an occasion piece for his metaphysical views, since not two pages later (32c–33a) he achieves his surface aim of showing that the world is unique, on grounds completely

---

[4] Owen (1).

[5] Nehamas has claimed that the second half of the *Phaedrus*, the discussion of rhetoric (257c ff.), is a rejection of the dualistic metaphysics of Forms presented in Socrates' second speech, the so-called "Great Speech" (243e–257b). Nehamas (2) xxxvii–xlvii. But Nehamas admits that the discussion of collection and of relations between Forms at 265d–266b in the second half is a backwards reference to the discussion of collection as recollection of Forms at 249c in the Great Speech. Nehamas (2) 64n149. But then all the discussions of relations between Forms in the second half of the dialogue refer back to and presume the dualist doctrines of the first half, since the discussions of relations between Forms in the second half compound upon and reference each other (265d–266b, 270a with 270d–271a, 273e, 277b), and so the Nehamas thesis on the (dis)unity and anti-metaphysical thrust of the dialogue implodes.

independent from and much less contentious than those found at 30c–31b. There, he simply argues that since the Demiurge did not leave behind any materials unused in his craftings out of which another world might be formed, the world which he did form is necessarily unique.[6]

To repeat then: my contention is that since Plato deploys his technical metaphysical vocabulary in this passage, a construal of the Platonic metaphysics which makes sense of his argument here, has a certain claim to be considered the correct interpretation of that metaphysics.

II. INTERPRETATION: THE DEMIURGE AND HIS MODEL

I will sketch the whole of what I take to be the correct interpretation of the unique world argument and then will argue separately for the various components of the interpretation (sections IV–VIII).

When Plato speaks of Forms as paradigms (παραδείγματα), I suggest that he means for them to be viewed as standards or measures under a certain description.[7] The unique world argument will help us determine exactly what that description is; but in any case, if Forms are standards, then we need to make a distinction among their various formal, external or metaphysical properties, that is, the properties which they each have

---

[6] The text: "He who put it together made it consist of all the fire and water and air and earth, leaving no part or power of any one of them outside . . . that it might be single, nothing being left over, out of which such another might come into being" (*Timaeus* 32c–33a, Cornford).

This passage clearly shows that Plato did not entertain the possibility that two distinct entities could have all the same parts, that, for instance, both the Elks Club of Smallsville and the Republican Party of Smallsville might each contain all the citizens of Smallsville.

[7] In the late dialogues "measure" (μέτρον) and its Platonic equivalent "limit" (πέρας) tend to replace "paradigm" (παράδειγμα) as the primary designation for the aspect of Forms as exemplars. Plato, in technical passages in the late group, begins to use the term "paradigm" to mean "parallel case" rather than "exemplar." Thus in the *Statesman,* the Form of weaving is said to be a paradigm for the Form of statecraft, since the two Forms have closely analogical structures (278e). There is no suggestion that statecraft is being viewed as an instance or example of weaving viewed as an exemplar. When "paradigm" is used in the sense "a parallel case," a "paradigmatic" relation will, of course, be a symmetric relation. It would be a serious mistake, though, to suppose that therefore Plato either intends the relation of exemplar to example to be a symmetric relation, or that he is always signaling a symmetric relation when he refers to paradigmatic relations. In the *Sophist,* the term is used to mean "parallel case" at 221b5, 226c11, and 233d3.

just in virtue of being Forms. There will be, on the one hand, what I will call their functional properties. These are the properties which a Form-standard has as an instrument fulfilling a role in Plato's scheme of things. These properties will primarily be epistemological. Forms have two epistemological roles. First, they in themselves are in some sense objects of knowledge (*Republic* 477a, *Timaeus* 28a, 52a) and second, they as standards or measures allow us to identify, by reference to them, the types and kinds of other things (e.g., *Euthyphro* 6e4). Thus by reference to The Standard Meter Stick, I can identify other things as being a meter long. Further, and perhaps derivatively from this second role, Forms as standards serve as models for the practical projects, both productive and moral, of rational agents (*Republic* VI 500e3, VII 540a9).

On the other hand, Forms as standards will have among their formal properties what I will call standard-establishing properties. These will be the properties requisite to a standard's ability to serve as a standard, that is, requisite to it having the functional properties it has. Most notable among these properties is some sort of permanence. A standard which can change with respect to that in virtue of which other things are identified by reference to it is not a (good) standard. I will eventually give an argument to suggest that Plato takes uniqueness to be one of the standard-establishing properties of the Ideas (section IV).

Now, I further suggest that for epistemological reasons which I will discuss (section III), the Demiurge supposes that the world can be improved if only he can somehow make it come to possess standards in some sense. These created standards immanent in the phenomenal realm will be derivative from the Ideal standards. In order to introduce standards into the world, the Demiurge will have to do two sorts of things. First, he will take over from the phenomenal flux an instance or image of a Form, an instance which he wishes to make into a standard instance or paradigm case. He will try the best he can to make this instance correspond in content to the standard of which it is an instance. He will try to reproduce as accurately as possible the Form's internal or proper attributes, that is, those things about each Form in virtue of which each is the particular Form it is. And he does this by eliminating any degree to which the instance might on its own fall short of or exceed its measure or standard. The Demiurge, that is, eliminates excess and deficiency from particulars and brings them into accord with their standards. So immanent standards will be 'perfect' particulars or 'perfect' instances. By "perfect particular" I mean a particular which corresponds precisely to the Form of which it is an instance. If I stretch a rubber band from its relaxed length to the length of the room, then at exactly one point in the stretching process the rubber band will necessarily correspond precisely in length to The Standard Meter Stick in Paris; at that one point it will be a perfect particular of that standard. At other lengths the rubber band is a degenerate instance of the standard; it falls away from the standard as being greater or less than its measure.[8]

Now, all immanent standards will be perfect particulars, but not all perfect particulars will serve as standards. For, second, the Demiurge, aside from making the standard case to-be a perfect particular, also will have to invest it with certain standard-establishing properties in order to enhance the prospects of the instance serving adequately as a standard case. Fixedness or stability is one such property which improves a particular's aptness to serve as a standard. A rubber band even though it is stretched to exactly the length of The Standard Meter Stick will not make a useful stand-in for The Standard Meter Stick as a measure of length; for it, on its own, lacks stability and will easily change its length (as we know by continuing to measure it under different conditions against The Standard Meter Stick). So too, if the pound measure in my shop is made of dry ice, it will change its weight as it sublimes, and so it will serve less well as a standard than a weight made, say, of iron. We can always tell of these derivative stand-in standards that they have changed, by comparing them to the one original standard for their kind (*The* Standard Meter, *The* Standard Pound).

In reproducing the standard-establishing properties of the Ideas, the Demiurge's project may be thought of as trying to introduce some of the Ideality of the Ideas into the phenomenal realm. But the Demiurge's abilities to perform this task are highly restricted. What the Demiurge can do is limited to a large extent by the way the world is given to him. He is not omnipotent like the Christian God. He does not invent his materials, he discovers them. And in his materials there is a certain cussedness,

---

[8] I suggest that Plato draws in a general way the distinction between perfect instances and degenerate instances of a standard or model in the *Sophist* (235b–236c), where he distinguishes between two kinds of imitation (μίμησις, 235d1). There is imitation which produces images (εἰκούς). These reproduce the exact proportions of their originals (παραδείγματα). And there is imitation which produces phantasms (φαντάσματα). These, though somewhat similar to the content of their originals, deviate from corresponding precisely to them.

Elsewhere in the late dialogues, we find the objective sources of the acceptable physical pleasures, the pure pleasures of sense, of the *Philebus* depicted as perfect particulars (51c). These objects include circles made with a lathe and straight lines drawn with a carpenter's rule. These figures are not characterized by being more or less than anything else—rounder or less round than *x*, straighter or less straight than *y*; they are not relative to anything else (πρός τι), their character is independent (καθ' αὑτα) of sliding scales of degrees.

It also seems that the "immanent characters" of the *Phaedo* (103b, e) are perfect particulars of the Forms of which they are instances. See Nehamas (1) 105–17, for a general defense of the presence of perfect particulars of Forms in the middle dialogues. For the immanent characters of the *Phaedo*, see especially 116.

which can partially defeat the enactment of his fully rational creative desires. The Demiurge can only make the world as like *as is possible* to the Ideal model (38b8–c1, 39e1).

In some cases, it will be impossible for him to introduce the same standard-establishing properties which the Ideas have. Thus, if, as I think is the case, non-corporeality is one of the standard-establishing properties of the Ideas, this will be a property which the Demiurge will not be able to introduce into the phenomenal realm to *any* degree. In other cases, while he will not be able to confer upon the standard cases exactly the same standard-establishing property which the Ideas have, he will nonetheless be able to provide a passable substitute. Thus, for instance, though the Demiurge cannot provide standard cases with the sort of permanence which the Ideas have, nonetheless he can provide them an ersatz permanence (so explicitly *Timaeus* 37d3–4) which serves in a pinch to provide the constancy which standards require in order to be standards. Finally, in the case of uniqueness as a standard-establishing property, it turns out for a number of reasons (see section VIII) that it too usually cannot be introduced into the phenomenal realm. However, over a very narrow range of kinds of things, this property can be introduced into the phenomenal realm to make a standard case. This is possible in the case of the Idea of living-thing, exactly because, as we will see, Plato has such peculiar views about the nature of living things (sections VI and VIII). But given these peculiar views and given the nature of the Demiurge's project, the world is made a unique living-thing-in-general so that it may serve as an immanent standard.[9]

---

[9] Parry's interpretation of the argument takes the uniqueness which the Demiurge copies to be an internal property, indeed an essential attribute of the Idea of animal. The uniqueness involved is taken to be the sort of uniqueness achieved when a whole or system is complete in containing as constituent parts all the possible instances of some type. Thus the Idea of animal is construed to be a system exhaustively including all possible types of Ideas of specific animals as its constitutive parts. This sort of uniqueness, the interpretation alleges, is imaged in the world by the Demiurge making the world a system which exhaustively includes, as constitutive parts, all animals. This reading has a number of interpretative difficulties: 1) the Demiurge could achieve this sort of uniqueness for the world without bothering to look to an Idea as a model, thus rendering the presence of the Ideal theory in the argument supererogatory; 2) as we have already seen the Demiurge indeed does produce, in the case of the material elements, this sort of uniqueness independently of the Ideas, rendering the current argument, so interpreted, otiose; 3) when the Demiurge does introduce the other animals into the world, this is viewed as a project quite independent of the unique world argument (39e3–5); the ἄλλα of 39e3 minimally includes the earlier project of making the world unique; therefore the uniqueness of the world as an animal-in-general is claimed to be established independently of the introduction of types

If the Demiurge's project is, as I have suggested, to introduce standards into the phenomenal realm by jointly producing there accurate images and standard-establishing properties, then Plato avoids a number of serious charges which critics have standardly laid against the unique world argument.

First, Plato and his Demiurge are not conceptually confused about the nature of the Demiurge's project. Specifically they have not confused the production of properties the presence of which in a thing makes it inherently good with the production of properties that make a thing an accurate, faithful or correct image of an original.[10] Rather, Plato and his Demiurge realize that for an image to serve as an immanent standard it must have two sorts of properties. On the one hand, to serve as an immanent standard for the type of things for which it is to be the standard, a phenomenal image needs to reproduce accurately that about its original that make its original the original it is. The Demiurge will try to reproduce accurately the proper or internal attributes of the original in the image which is to serve as a standard. If we were to pick or produce a painting of Winston Churchill with the aid of which we were going to proceed to identify other paintings as being paintings of Winston Churchill, we would want our standard painting to image as accurately as possible the distinctive features of Winston Churchill. On the other hand, for an image to be adapted for use as a standard, it will also need those standard-establishing properties which make it possible for, and enhance the aptness of, the image to serve as a standard. These will typically include properties like orderliness, stability, and unity.

Now it is not hard to understand that these properties (orderliness, stability, unity) might be misconstrued by critics as being, just in themselves, final constituents of Platonic goodness. Neoplatonists, critics who have a right-wing political bent or who think Plato has such a bent, and the aesthetically-minded are all likely to suppose that these properties of order, stability, and unity just are what it means to be good for Plato and

---

and tokens of other animals; 4) Plato simply nowhere characterizes either essentially or accidentally the unique World-Animal as a system of other animals. Now, such composite creatures do exist. Individual lichens, for instance, are composite organisms made up totally of a fungus and an alga living symbiotically. And the bizarre figure with which Plato ends Book IX of the *Republic* (588c–589b) shows that the notion of composite creatures is not foreign to him. Plato could easily have described the World-Animal as a composite creature, were he so inclined.

[10] This charge of conceptual confusion is laid by Patterson (1) 116–19. On Johansen's reading of the unique world argument, the workings of the Demiurge are confused in exactly the way Patterson detects, but oddly Johansen does not see the confusion as a problem. Johansen 54.

that the Demiurge needs no further justification for their introduction into the world. Thus these critics will claim that if the Demiurge should make any improvements in the world at all, he will make these improvements (order, stability, unity). Such a critic is Cornford, who in speaking of the unique world argument declares: "Uniqueness is a perfection, and the world is better for possessing it."[11] From this perspective of viewing uniqueness, stability and the like as intrinsic goods or perfections, it will naturally enough appear that the Demiurge is involved in two wholly unrelated projects: 1) the making of accurate images and 2) the introduction of properties which count as intrinsic goods. The correct reading, though, is to view uniqueness, stability and the like as having instrumental value, as making possible and enhancing the ability of images to serve as standards. The making of accurate images and the introducing of standard-establishing properties are joint parts of the Demiurge's project.

Plato's unique world argument has also been accused of constituting a gross fallacy of division.[12] According to this charge, the Demiurge allegedly supposes that *since* the world is (to be) an instance of a Form, and *since* this Form is an instance of uniqueness, therefore the world is (to be) unique for *these* reasons. However, that some phenomenal objects are to serve as standards (of a sort) motivates their coming to possess the standard-establishing properties of the Ideas, including uniqueness. This motivation is quite *independent* of these phenomenal objects simply being instances of the Forms of which they indeed are instances. Therefore, the Demiurge's thought is not riddled with a fallacy of division.

The Demiurge has also been accused of being quite mad, since he at least seems to be indiscriminately reproducing in his copy each and every property of his original. Such an indiscriminate reproduction of properties would indeed be a crazy way of proceeding for a number of reasons. First, sometimes an original will possess properties which are irrelevant to its serving as an original (for example, having dents and scratches). One would not indiscriminately reproduce these in a copy. Second, there are some properties of an original which, though relevant to the original serving as an original, are wholly inappropriate for the product for which the model is used. Thus it is entirely appropriate that

---

[11] Cornford (2) 43. Zeyl, Broadie, and Johansen basically follow Cornford here and take the function of the Demiurge to be the introduction of intrinsic goods or perfections into the phenomenal realm. Zeyl: the Demiurge "imposed order and beauty upon a preexisting chaos," turning the world into "a work of art." Zeyl (2) xiii, xiv. Broadie: "Beautiful animal life is the ultimate cosmic objective." Broadie (2) 177. Johansen: "Making the cosmos share the goodness and beauty of its model is the whole point of the creation." Johansen 54.

[12] This charge and the next are laid by Keyt (1).

an architect's blueprint should be made of waxy paper, but one would not make a house of waxy paper because its plans were made of such paper. Third, an original will possess some properties which are logically impossible to reproduce in a copy, for example, the age of the original and the numerical self-identity of the original.

On my reading of the Demiurge, he is not subject to these charges. He does not indiscriminately reproduce the properties of his model, rather his project of making immanent standards provides him with a principle for selecting which of the attributes of his original are to be reproduced. It only *appears* that the Demiurge is picking and reproducing properties indiscriminately because so many of the attributes of the Ideas are standard-establishing properties and he will try to reproduce all of these. The Demiurge's choices are relevant and appropriate, guided by the nature of his specific project.

The Demiurge is therefore not wanting in intelligence, ingenuity, or sanity. I suggest that the figure of the Demiurge is a coherent one. Reasonable attacks upon the Demiurge will not take the form of accusations of absurdity, but will have to take the form of claims of philosophical economy, that is, will take the form of trying to show that what Plato wishes to explain by the figure of the Demiurge, namely, our possession of certain cognitive capacities (section III), can be achieved with considerably less conspicuous metaphysical consumption, just as the gods of the cosmological or first cause arguments are perhaps shown to be otiose by Newtonian mechanics.

III. IMMANENT STANDARDS AND DEMIURGIC INTENT

Textual evidence that Plato did believe that there are such things as standards immanent in the phenomenal realm and that he did not suppose that only Forms can serve in a way as standards comes from several diverse sources.

Within the *Timaeus* itself, the distinction between transcendent standards and immanent standards is drawn in the very opening section of Timaeus' discourse where he is laying out his principles. Here it is claimed that there are two kinds of models, standards, or paradigms (παραδείγματα, 28c6–29a2). One kind is called "eternal," a periphrastic reference to the Ideas. And, in contrast, one kind is called "generated" (γεγονός), by which I take Plato, in virtue of the contrast, to mean nothing more than "is part of the phenomenal world of becoming."[13] Later in the *Timaeus* Plato makes it clear that he intends the Demiurge's making of time to be viewed as a production of such sensible standards (37c–38e).

[13] For more discussion of these generated paradigms, see section IXA below.

This is particularly important for my interpretation; for Plato explicitly claims that the making of time and the making of the world into a unique animal are parallel, analogous projects of image making. The sense in which each of the two replicas is a likeness of that of which each is an image is the same sense (39e3–4); indeed the production of time is seen as paradigmatic for all of the Demiurge's craftings.

When Plato says that the Demiurge makes time he does not mean that the Demiurge creates temporal succession. The Demiurge does not, that is, create events as having a determinate, transitive, asymmetrical, irreflexive order. Rather the Demiurge makes the means of measuring such orderings against a standard (for details, see chapter 2). The Demiurge makes a clock (or more precisely makes clocks) by which we *tell time*, that is, by which we measure that which is measurable about events and durations. The time which the Demiurge creates is a standard or measure immanent in the phenomenal realm, a standard which images an Ideal standard (37d5, 7; 38a7; 38b8–c1; 39e1–2).

Immanent demiurgically generated standards or paradigms are also to be found in the *Republic.* In the central books, we are told that the ordered parts of the phenomenal realm ("the embroidery of the sky") serve as παραδείγματα (529d7–8) in the study of the special theoretical sciences (astronomy in particular). It is clear from the context that the meaning of the term παραδείγματα here falls within the range of its senses which cluster around "exemplar" rather than around "example" or even "parallel case." For these immanent paradigms seem to be treated like blueprints in nature and function: "It is just as if one came upon plans carefully drawn and executed by the sculptor Daedalus or some other craftsman (δημιουργός) or artist" (529d8–e3). Plato in his analogy here is referring to the custom of Greek sculptors, in making their final products (say, marble horses), to work from models or paradigms which themselves are artificial objects (clay horses) rather than to work directly from models which are natural objects (living horses).[14]

The paradigms which make up the embroidery of the sky are explicitly said to be the products of a demiurge (530a6, cf. 507c6–7), and are explicitly said to be material and visible (σῶμά τε ἔχοντα καὶ ὁρώμενα, 530b3). Some of these earthly paradigms turn out, as in the *Timaeus,* to be the parts of time: "days," "nights," "months," "years," and "*other* stars" (530a7–8), by which designations it is clear, especially given the context, that Plato is referring to the heavenly bodies as measures or standards rather than as measurements taken from standards. Indeed the discussion of time in the *Timaeus* is basically just an elaboration of the brief

[14] For a discussion of the role of models in the production of Greek sculpture, see Bluemel 36–39. Note illustration 28, in which the divine craftsman Athena is represented as making a clay model which a marble sculptor will then use.

claims about the stars and planets serving as visible paradigms in the *Republic*.

That Plato believed that some instances of Forms are standards or measures is most clearly and persistently articulated in the *Philebus*. In it we find both transcendental measures (25d3 BT; 26d9, πέρατος; 57d2; 66a6–7) and measures immanent in the world (26a3, 26d9, 66b1–2).[15] The immanent measures of the *Philebus* are phenomena which correspond exactly to a Form and which have had removed from them by demiurgic activity the propensity to slide along scales of degrees (to become hotter and colder, wetter and drier, thicker and thinner, to move higher and lower, more quickly or more slowly, more forcefully or more gently, etc.), or in general to be subject to the more and the less. When demiurgic making fixes a phenomenal object at a determinate, quantified point on a sliding scale of degrees (say, a fever- and chill-beset human body at 98.6° F), excess and deficiency and the tendency to exceed or fall short of measure are all destroyed (24b1–2, 24d1–2, d2–3).[16]

The introduction by the Demiurge of measures or standards into the phenomenal realm is extremely important for Plato's epistemology. For it means that if one has only the phenomenal realm (and not also the Ideal realm) as the object of one's cognition, nevertheless one is not limited to making judgments based entirely on merely relative comparison or merely relative measurement. By "merely relative measurement" I mean that ability which Plato claims we have, usually by direct sensory

---

[15] In the *Philebus* the paronymous forms of μέτρον, namely, ἔμμετρον and σύμμετρον, generally refer to immanent rather than transcendent measures or standards (26a7, 8; 52c4; 64d9; 65d10). They should be translated respectively not, as they usually are, as "moderate" and "symmetrical," but, with reference to their etymological roots, as "affected with measure" and "with measure." The issue of the status of measures in the *Philebus* is much debated. The chief error that anti-dualist interpreters of the *Philebus* make is to move from the correct observation that some of the measure of the *Philebus* are phenomenal, material objects to the false generalization that all the measures in the *Philebus* are phenomenal, material objects. For defenses that some measures in the *Philebus* (and *Statesman*) are Platonic Forms, see Mohr (1), (2), and chapter 15.

[16] That the-more-and-the-less is destroyed, ceases to exist, in the process of demiurgic making shows that Plato's use of "mixture" as a description of the process of something coming into being from measure and the-more-and-the-less is highly metaphoric (e.g., *Philebus* 23d1, 7, 25b5). For in a mixture (say, of water and wine that are mixed to make the wine potable), the 'elements' of the mixture (water and alcohol) are both preserved in the mixture. I suggest that in the *Philebus* the mixture metaphor is used by Plato to mark only the ontological independence of the components *prior* to the process of making. It certainly does not entail that all measures have been dragged down into the phenomenal realm.

inspection and in any case without reference to measures or standards, to say that, on some scale of degrees, one thing is greater than another, or even possesses the same degree as another, but which does not entail the further ability to say what degree on the relevant scale of degrees either thing possesses. We would have the latter ability only if we could appeal to a standard or measure (*Statesman* 283c–285a). When we are provided with immanent standards, we have the improved epistemological skill of being able to make precise identifications of phenomenal kinds, without having to make appeals (at least directly) to anything other than the phenomena.

As an illustrative example of what Plato means by merely relative measure and measure against a standard, let us take Plato's own example of (celestial) clocks. The Demiurge in creating clocks is not (as mentioned) creating temporal succession. Temporal succession exists whether clocks exist or not. Further Plato is not claiming that in an acosmic world in which there are no clocks we would (should we exist) be unable to perceive and make judgments about temporal succession. In a world without clocks we still can make merely relative measurements of time, that is, raw judgments of earlier and later, without appeal to a clock as a standard, just as we can make merely relative measures of length, weight, and the like, without appeals to standards for those dimensions. Thus, if I snapped the fingers of one hand, paused, and then snapped the fingers of the other, anyone in the room who was paying attention could say which of the two events occurred first *without* appealing to a clock. But the two events could also be given a temporal ordering *by* appealing to a clock as a standard of measure for dating events. Each snap could be given a date by reference to a clock, say, 6:31 and 6:32 and given a temporal ordering as the result of this matching of each event severally to parts of the clock and a comparing of the parts rather than as a result of comparing the two events directly against each other (for details, see chapter 2, section IIC).

In general, the presence of immanent standards allows us to identify individuals correctly and precisely. Derivatively, immanent measures allow us to make precise comparisons between things measured; for we are able to compare their determinate measures. Without appeals to standards or measures we are only able to say that one thing is greater or less than another but without being able to say what either thing determinately is.

In the *Philebus* Plato spells out more specifically what he takes to be the cognitive and practical functions of immanent standards or measures. They are the objects of true opinion and of the applied arts and crafts (like accounting) (*Philebus* 55d–e, 62b–c, 66b1–2), in the same manner in which the transcendental measures are the objects of reason and the purely theoretic sciences (like number theory) (57d2, 58d6–7, 61d–e,

66a–b). False opinion has as its object not nothing nor even what is not the case, but rather the phenomena which fall away from standards, phenomena, that is, which are subject to the more and the less and which on their own are approached only by merely relative measurement. Such objects tend to cause errors of judgment and practice, when we have only them before us and so are unable to identify and assess them with reference to standards (so also *Republic* 584a, 586b–c). This view of false opinion can be more easily understood, if it is remembered that opinion is for Plato at least as much a matter of pragmatics as of semantics. Opinion is more a matter of determining how we get along in the world than a determination of propositional accuracy. Even true opinion is to a large degree a matter of forecast (*Philebus* 55e5–56a1, 62c1–2). True opinion is what allows us to succeed at our projects in the world, by intelligently guessing correctly the right course of action (*Republic* VI, 506c). Possessing models or blueprints for identification and guidance is of course very useful, maybe even necessary, in this state of affairs (62b5–9; ψευδοῦς χανόνος = immanent standard; see *LSJ*, s.v. κανών, II). We have a better grade of cognition if we have access to standards and measures than if we do not. Without numbering, measuring, weighing and the like that immanent measures and measurings make possible, we are left only with unintelligent guesswork (εἰκάζειν), to working the senses by trial (ἐμπειρίᾳ) and by knack or rule of thumb (τριβῇ), and the ability to make lucky shots (*Philebus* 55e–56a). By contrast, the builder's trade, because it uses lots of immanent measures and standards (the straight edge, the peg-and-cord, the compass, the plummet, and set square), is more 'scientific' (τεχνικωτέραν) and achieves exactness (56b–c).[17]

[17] The *Philebus'* use of εἰκάζειν here helps clear up a problem from the Divided Line in the *Republic*. The lowest level of the Divided Line consisted of images of natural objects (trees, cows) and everyday manufactured objects (509e–510a) and yet these objects of the lowest level of the Line, especially when cross-compared to the shadows of the Cave Analogy, are meant to represent the *usual* objects of our everyday cognition—a claim which even in the era of television and cyberspace seems wildly false. See Annas (2) chapter 10. Rather the things that are their models (natural and artificial substances) are the usual objects of our everyday cognition. Now, the mode of cognition of the lowest level of the Line is called εἰκάζειν, whose range of senses include not just "conjecture" but also "comparing one thing with another" (see *LSJ*, s.v.). Εἰκάζειν is also the mode of cognition for some objects within the phenomenal world in the *Philebus*, a mode that *Philebus* assigns to people who do not have access to even immanent standards and measures. The lowest two levels of the Divided Line then can be translated or mapped onto the *Philebus'* division of objects and cognitions within the phenomenal world: some things, most things, are without measure and are only subject to conjecture and to merely relative comparisons, comparisons which leave us unable to identify anything precisely, while other things, far fewer, are

The presence of measures and standards in the phenomenal realm then benefits that realm by largely constituting its intelligibility. For they allow us to make accurate and useful identifications when we have only the phenomenal realm before us. But they do something more, indeed for Plato something more important even than this. More so than randomly selected phenomena ever could, immanent standards also help get us on the road to full knowledge and philosophy. The usefulness of artificial models is not restricted to their serving simply *qua* paradigms, that is, making possible the identification or production of further particulars by reference to them. They are also useful in coming to understand the models of which they themselves are instances.

Take again Daedalus' architectural plans, the plans to which the orderly heavens are compared in *Republic* VII (529d8–e3). A blueprint is not only useful in making houses, but also useful in coming to understand *what it is to be* a house, the definition of a house. When condominiums were first becoming popular in America, real estate brokers frequently included blueprints in advertisements for them. The function of the blueprint in the advertisement was to convince the buyer that what she was buying was a house. Immanent standards turn us around toward the original of which they themselves are images and in a more focused way than a randomly selected image could. This refocusing and redirecting of the mind is part of the reason that the special theoretical sciences of *Republic* VII lead us toward dialectics and the Forms.

The details are thin, but this refocusing seems to be the chief function that Plato has in mind for the World-Soul and the cosmic chronometers in his treatment of them as immanent standards in the *Timaeus*. The grand conclusion of the first half of the *Timaeus*, the half devoted to articulating the works of reason, is that the study of the heavenly chronometers leads to the invention of number and from there to philosophy itself (47a–b). Apparently, as in the *Republic* VII, the study of numbers leads, if not automatically, at least naturally, to the study and understanding of Forms (524e–525a, 526a–b). And the immanent paradigm that is the rational World-Soul serves as a model for our general rational capacities:

> The god invented and gave us vision in order that we might observe the circuits of intelligence in the heaven and profit by them for the revolutions of our own thought, which are akin to them, though ours be troubled and they are unperturbed; and that by learning to know them and acquiring the power to compute them rightly according to nature, we might reproduce the perfectly unerring revolutions of the god [i.e., the World-Soul] and reduce to settled order the wandering motions in ourselves. (47b–c, Cornford)

---

immanent measures, which provide objects for true opinions and allow us to make precise identifications. With the objects of the lowest level of the Divided Line so translated, the problem from the *Republic* evaporates.

With our rational capacities set as the World-Soul's are, we are able to make the same sort of true judgments that it does about both the phenomenal world and the world of Forms (37b–c). Time, as immanent standard, provides us the mental content needed to get us on the way to Forms; cosmic reason, as an immanent standard, provides us the mental capacities to do so. Jointly they redirect us toward the models after which they are made.

I suggest that these sorts of benefits are the primary aim of the Demiurge's craftings. The improvements which he works on the phenomena are primarily epistemological rather than aesthetic. He is not primarily trying to make the world look better, to make it prettier.[18] Even though he is a craftsman who uses models, he is not primarily an artist, a producer of beauty. Rather his craftings are largely deployed to introduce into the world, if possible, those properties which in turn make it possible for phenomenal particulars to serve as standards.

If my reading of the Demiurge's aims and practices is anywhere near correct, the figure of the Demiurge of the *Timaeus* is a singular accomplishment in the history of ideas. We do not find a figure with his function and nature in either prior or subsequent theological speculation. The strongest traditions of theistic speculation in the West have treated the divine either as primarily a source of motion or as primarily a source of order. The one tradition can be traced back to Empedocles, whose Love and Strife set the world's components in motion. This tradition blossoms in the cosmological or first cause arguments for the existence of God. The other tradition can be traced back to Anaxagoras, whose divine Mind set the world's components in order. This tradition blossoms in the teleological arguments for the existence of God.

If my interpretation of Plato's Demiurge is correct, the Demiurge does not fit into either of these traditions. For, on the one hand, far from being primarily a source of motion, the Demiurge is actually a source of stability and permanence, since these are conditions which enhance the prospects of some phenomena serving as standard cases. This is not to deny that the Demiurge might be thought of as necessarily moving things around: a forge worker will, after all, move things around in his project of tempering a piece of steel. But, it does mean that the Demiurge is not hypothesized to explain the origin or perpetual occurrence of motion in the universe, as gods are called upon to do in cosmological argu-

[18] The polyvalent term καλός at *Timaeus* 29a2, used to describe the quality with which the Demiurge invests the world, might be translated as "fine." Plato is looking for a non-moral normative term with which to contrast the moral goodness, the good intentions, of the Demiurge himself in the same clause (ἀγαθός, a3). Even translating καλός here as "good," as Cornford does, is preferable to narrowly translating it as "beautiful," as Zeyl does. Zeyl (2) 14.

ments, including Plato's own version of those arguments in *Laws* X (894e–895b).

Further, on the other hand, though the Demiurge may be viewed as a source of order, regularity, and other 'intrinsic' goods, neither is their production his final end nor is he hypothesized in order to explain their presence in the world as are hypothesized the gods of teleological arguments. The nature of the Demiurge is completely compatible with Plato's claim that there are "traces" of order even in the pre-cosmic era (53b2).[19] The Demiurge is not responsible for *all* regularity and order. He is responsible only for those final forms of order which proximately serve functions and manifest purpose. Such final orderings may be entirely new formations, but they may equally well be, to a large degree, incorporations of pre-existing orderly elements which the Demiurge simply appropriates without alteration from the pre-cosmos (see *Timaeus* 46c–47a).[20] Prior to their appropriation such elements will not correctly be said to manifest purpose, afterwards they will. But whether a demiurgic ordering takes the form of an appropriation or a formation, it is carried out not as a final end on its own but as part of the Demiurge's project of making immanent standards.

If the Demiurge bears a resemblance to any of the gods of the Western theistic traditions, it is to the god of the ontological argument, or at least, of those variants of the argument, like that in Descartes' Third Meditation, which employ the principle of sufficient reason. For both the Demiurge and the god of this ontological argument are hypothesized as necessary beings in consequence of certain epistemological 'phenomena'. Though, even here there is substantial difference between the two. For the god of the (Cartesian) ontological argument is hypothesized as being necessary to explain sufficiently an agent's *actual* possession of a *particular* concept (e.g., my actual possession of the idea of infinity). But the Demiurge is hypothesized, on my account, as being necessary simply for establishing the conditions in virtue of which we *potentially* possess *all* concepts, i.e., possess the means of forming true judgments with regard to all types of things.

I now turn to providing justifications for the various components of my interpretation of the unique world argument. First and foremost, it must turn out that standards in their primary sense are unique.

---

[19] Vlastos thinks that Plato's commitment to some orderly elements in the pre-cosmos is fundamentally incoherent when placed against a metaphysical background which includes a demiurgic god. Vlastos (2) 389–90, (3) 413–14.

[20] And see Johansen's particularly fine discussion of Plato's accompanying causes. Johansen 99–116.

IV. UNIQUENESS AS A STANDARD-ESTABLISHING PROPERTY

If we are to make predications or identifications of kinds by reference to standards, and not by means of merely relative measurement, then the standard for any given kind of measurement must turn out to be unique and consequently turn out simply to be a given rather than something which is explained.

For, if two things are such that they may both equally serve as standards for the same type of thing, they would have to be numerically distinct but formally similar to each other, otherwise we would not know that they were of equal service as standards for the same type of thing. But the condition of similarity or formal identity between them destroys the very possibility that either of them is a standard. For the property which they have in common and which allegedly determines each as the standard it is can only be said of them severally by reference to a third thing which neither of them is, if *ex hypothesi* all identifications are made by reference to standards and *not by merely relative comparisons.* And so it, rather than they, would be the standard for their kind. They would be mere instances of it. Standards *qua* standards are, therefore, necessarily unique for each sort of thing measured. Plato knew this and the argument here, I suggest, is how the *reductio* argument of *Republic* X against multiple Forms of the same type is to be read (597c), as should also be read the argument for the uniqueness of the Idea of animal in our passage (*Timaeus* 31a4–7). The very conditions which allow for making similar determinate predications prevent there being multiple standards of the same type.

It is true that we colloquially speak of multiple standards for a given kind ("The sewing teacher had a tape-measure for each student," "The pawnbroker would not take in any more pocket watches"). We make these judgments of plurality, though, only as the result of merely relative comparisons. Suppose I have two balsa wood sticks. On the one hand, I may try to claim of them that they 1) are both standards and 2) are the same length. But claim 2) can be made only a) in virtue of merely relative measure *or* b) in virtue of referring the sticks severally to some third thing. Alternative b) destroys the sticks as independent standards, since it makes them instances of a standard for their kind; and alternative a) fails to offer any basis for claim 1), namely, that they are standards. We are therefore lacking any reason to suppose that they are each a standard and a standard for the same type of thing. On the other hand, I may stipulate one of the two sticks to be the standard of length, but in that case the other stick has the length it has in virtue of the relation it holds to the stipulated standard and so ceases to be itself a standard (at least in the same sense as the other). This is quite obviously the case if I take my stipulated balsa wood meter stick and go measure The Standard Meter Stick in Paris with it. The Standard Meter Stick in this case is reduced to being treated merely as any other material object and retains the name Standard only by a quirk of linguistic history and not because of its nature. Only derivative standards (which in principle can be checked against an independent standard for constancy and clarity of content) are capable of multiplication. Standards just by themselves are unique for each type of thing measured.

There is an irreducible givenness about standards which goes hand-in-hand with their being unique and which eliminates the possibility of making significant predications about them with regard to their being the standards that they are. Standards cannot be defined as solutions or instances of sets of conditions, conditions a) which, as it were, give the "essence" of the standard, b) which may be instantiated any number of times, and c) which are spoken univocally, when predicated of both a standard and its instances.[21] For conditions which establish a standard for some dimension (weight, length, etc.) invariably must presuppose fixed, constant units in the very dimension for which they are being used to try to establish the measure, but this constancy could only be determinately established by reference to the very standard which the set of conditions is being used to try to establish. By "determinately established" I mean "determined without resort to merely relative measure." Attempts to define standards as solutions of sets of conditions which avoid appeals to merely relative measure turn out to be hopelessly circular, presupposing the very standard which they seek to establish. Thus, for example, attempts to define *a* standard (and a *repeatable* standard) meter as a certain percentage of the polar circumference of the Earth presuppose the size of the Earth to be fixed, something which could only be established determinately by the use of a standard measure of distance. Similarly, trying to define *a* standard repeatable meter as the length of a pendulum of a certain period of swing presupposes the length of the pendulum to be constant through the period, again something which could be established determinately only by reference to the standard

---

[21] For an attempt to view standards as instances of sets of defining conditions, see Strang. When applied to the Platonic Ideas, this view of standards will instantly generate the logical viciousness of the Third Man Argument. This view of standards is, at a general level, an attempt to make the existence of standards compatible with empiricist and behaviorist epistemologies. But it is specifically motivated, I suggest, by a conceptual confusion traceable to Wittgenstein, namely, the assumption that standards are merely means or instruments of representation but are not also that which gets represented. Wittgenstein (2) section 50. On this view, what gets represented in the instances of a standard is the very conditions by which the standard itself is defined. Against this conceptual confusion, see chapter 12, section III.

measure of distance, which the set of conditions is trying to establish, but which in fact is presupposed by the conditions.

We need to distinguish between conditions which one might attempt to use to establish or define a standard for some dimension (length, weight, etc.) from conditions which one uses to try to maintain the stability of a material standard once the standard is established simply by stipulation. These latter conditions might include the regulation of position, temperature, air pressure. These conditions are standard-establishing properties which make a material standard a better standard, but they are not conditions which define the standard as the standard it is.

To return to the Demiurge: if the Demiurge is introducing standards into the world, albeit derivative or immanent standards, he will want to make each kind of standard unique as having this standard-establishing property. If there are no impediments to the Demiurge making the world unique, he will do so as part of his project of increasing the intelligibility of the world.

V. ENCOMPASSING IDEAS AND PARTS OF IDEAS

It could be argued against the interpretation I'm advancing that in the unique world argument the unity of the Idea of animal is pretty clearly intended to be a unity in the sense of a whole of essential constitutive parts and that this unity becomes unique if it exhaustively contains as its part all (and only) the instances of some kind.[22] On such a reading the uniqueness of the Idea of animal-in-general turns out to be an internal, proper attribute of the Idea, rather than a formal or metaphysical property of the Idea and so *a fortiori* is not a standard-establishing property of the Idea. That Plato intends uniqueness here to be interpreted as exhaustive completeness has been taken as conclusively established by the claims in the argument that the Idea of animal contains, encompasses, or embraces as its parts all of the other Ideas of animals, and so is unique for this reason (30c7–8, 31a4–5). On this reading of the argument, the world is made unique as allegedly exhaustively containing as essential parts all the instances and/or kinds of animals just as the Idea of animal is unique as exhaustively containing all the other Ideas of animals. This reading runs into serious difficulties when it tries to read this sense of uniqueness into the phenomenal realm.[23] Even more importantly, though, I will argue that this interpretation completely misconstrues the Platonic parts-of-Ideas doctrine. When understood correctly, the notions

[22] Parry and Patterson (1) construe uniqueness in this way; Zeyl endorses Parry's and Patterson's general reading of the argument. Zeyl (2) xxxvii–xxxviii.

[23] See n9 above.

"part of an Idea" and "encompassing Ideas" far from contradicting my suggestion that the Idea of animal is unique as being a standard (rather than as being an exhaustive whole) will actually point to the same conclusion. For I will be arguing that the relation 'part of' used in our passage is to be understood as the relation of instance-to-standard.

By parts of an Idea, parts which are themselves Ideas, Plato simply means those Ideas which merely have in common the property which the one Idea enables us to pick out among the others severally.[24] It is important to remember that Plato has no technical vocabulary for the notions universal and particular. In the absence of a developed technical vocabulary, Plato seems awkwardly to resort to pressing the language of 'wholes and parts' into service for the notions of the general and the particular, as indeed does even Aristotle in his use of the expression "that in virtue of a whole" (τὸ καθόλου) as his frozen formula for "universal."

To repeat: my suggestion is that for one Form to *contain* others as parts is just for the others to all be correctly called by the same name because of a relation that each holds to the one Form. However, the one Form is not reducible to the many parts. A 'general' Form is not a commutative universal; it is not simply equivalent to that which its instances severally but in common might possess. For Plato uses the language of containing, embracing, surrounding, and encompassing just exactly to claim that the thing which embraces cannot be reduced to that which it embraces but rather has a substantial existence independently of the things which it embraces, whether these things are merely similar things or whether they are components of a system or are contained in some other way.[25] This holds quite generally. Thus at *Timaeus* 34b, the substantial independence of the World-Soul from the World-Body is asserted by the claim that the Demiurge caused the world's soul "to extend throughout the world's body and wrapped the body round with soul on the outside"(cf. 36e). The language here of "extending through" and "encompassing on the outside" parallels precisely the language which Plato uses to describe the relations of 'general' Forms to their instantiations in other Forms at *Sophist* 253d. Here the point is that though the more 'general' Forms are in some sense present in more 'specific' Forms as extending through them (διατεταμένην, d6), nonetheless, the more 'general' Forms exist in-

---

[24] I do not believe that Forms are universals in the sense of things that other things have severally but in common. A Form allows us to identify the common properties among things which in fact share common properties, but it is not even the case that all the things that are correctly identified by reference to one and the same Form must have common properties among them. See chapter 12, section I.

[25] As in the obscure use of the term at 33b3, where it is said that the sphere "contains all of the other figures there are."

dependently of their parts as surrounding them on the outside (ἔξωθεν περιεχομένας, d8).

This independence of the more 'general' Idea from its part-instances is emphatically asserted at *Sophist* 250b. Here Plato says that Motion and Rest are embraced by Being, and explicitly elaborates this as meaning (n.b. ὡς): 1) that Being is a third thing *over and above* the other two (τρίτον τι παρὰ ταῦτα, emphatic by position) and 2) that they are embraced as its parts simply because they commune with or participate in it (κοινωνίαν) (250b7–10).

When, in the unique world argument, Plato says that the Idea of animal contains its parts "just as this world contains us and all the other creatures" (30c8–d1), I suggest that the analogical feature to which he is referring is simply that feature in his conception of "containing" by which the container stands over and above, as a separate entity, that which it contains. The Idea of animal is like the world in the limited respect that each stands over and above, as a separate entity, that which it contains.[26]

We find the two claims 1) that a 'general' Form stands independently of the instance-parts which it encompasses and 2) that the encompassed parts are merely similar things (as the result of participation in the same Form) stated formally and as general doctrine at *Statesman* 285a–b:

> When a person [who performs divisions] at first perceives the commonality (κοινωνίαν) of many things, he must not give up until he sees all the differences in the commonality insofar as the differences have the status of kinds (εἴδεσι), and conversely, when all sorts of dissimilarities are seen in a large number of things, he must not be discouraged or stop until he, by having enclosed (ἕρξας) all things which are related to each other (σύμπαντα τὰ οἰκεῖα) within one similarity, has encompassed (περιβάληται) them with a being of some type (γένους τινὸς οὐσίᾳ).

The "many things" in this passage are Forms or kinds (γένη, 285b8) which themselves are parts (μέρη, 285a7). The 'general' Form is not a whole or arrangement of parts; it is an entity which stands over and above parts which it encompasses and which are merely similar to each other.

Now, if we further wonder how the merely similar parts happen to be similar to each other and how they stand to the 'general' Form, Plato tells us the answer to these questions in the discussion of the parts of Difference in the *Sophist*:

---

[26] Parry mistakenly takes 30c8–d1 as decisive evidence that Plato is committed to viewing 'containing' and so also 'uniqueness' as internal properties of the Idea of animal. Parry 9.

> Each one [of the parts of the Idea of difference, cf. 257c] is different from the others not by reason of its own nature (διὰ τὴν αὑτοῦ φύσιν), but because of its participation in the Idea of difference (255e4–6).

Forms which are parts possess common features not because of what each Form is on its own, but because each participates in yet another Form. One interpreter of the unique world argument claims that "it is debatable in these dialogues [*Sophist, Statesman*] whether a Form is ever a part of any Form in which it participates."[27] Plato here is quite explicit, to the contrary, that one Form is part of another just *exactly because* it participates in the other.

VI. PLATONIC SPECIES

Now the objection might be raised that while some parts of Forms are mere instantiations, Plato clearly views some parts as having a much stronger relation to the 'general' Form. For at *Statesman* 262b–264b, Plato distinguishes parts of Ideas which are parts and nothing else, from parts of Ideas which are also "species" in a peculiarly Platonic sense. Platonic species are those more specific concepts which fulfill the two following conditions: 1) they are entailed by the more generic concept of which they are parts or divisions *and* 2) they make manifest some structural feature of the more generic concept. Thus Plato gives as illustrative analogies of parts which are not species barbarians and Hellenes, which are parts but not species of man, and the number 10,000 which is a part but not a species of number; whereas even and odd are both parts *and* species of number, and male and female are both parts *and* species of man (*Statesman* 262b–264b). Plato has picked his examples here with some care. The number example clearly indicates that condition 2), and not just condition 1), is what is at stake in his distinction, since the number 10,000 is in fact, as Plato was well aware, entailed by the Idea of number or The Natural Number Line.[28] And the example of man clearly shows that the converse of condition 1) is not part of the distinction Plato is making, that is, Plato is not saying that parts which are also Platonic species entail the more 'general' Idea of which they are parts. For Plato knew that many things besides humans have genders.

Does Plato take the parts of the Idea of animal to be more than mere parts? Does he intend them also to be "species" in his technical sense,

---

[27] Patterson (1) 107n3. Parry makes the same mistake: "Even if [it is] shown that the two living creatures participate in the Form *Zöon,* [it will not have been] shown that they are parts of *Zöon.*" Parry 8.

[28] See chapter 15.

that is, do the parts of Animal stand to Animal, for Plato, as even and odd stands to number? Some critics answer these questions affirmatively.[29] This answer unfortunately overlooks what Plato actually says about the nature of kinds of animals. For Plato, a living thing is just a soul in a body (*Phaedo* 105c9–11, *Timaeus* 30b4–5 with 7–8, *Sophist* 246e5–7). And I suggest that the Idea of living-thing is just the Idea of soul (so roughly, *Phaedo* 105c–106d), or more precisely, of soul as soul is directed to a certain environment. The various material environments in which soul finds itself will specify the types or 'species' of creatures which there are. The 'specific' Ideas of animals will be the Ideas of the primary bodies (earth, air, fire, water) (*Timaeus* 51b) in their aspect as participating in the Idea of animal or of soul-in-an-environment. And this is indeed how Plato describes (the 'essences' of) the 'species' of animals. The kinds of animals are nothing but souls, viewed as sufficient vivifying forces, in sorts of matter and derivatively in the sorts of natural regions (39e–40a) that arise from the natures of the four primary bodies (53a). Thus, gods or fire creatures are described essentially as souls in bodies made (mostly) of fire and residing in the fire region (40a). Fish or water creatures are souls in bodies made (mostly) of water and residing in the water region. Birds or air creatures are souls in bodies made (mostly) of air and residing in the air region. And solid, fleshy creatures are souls in bodies made (mostly) of earth and residing on earth (cf. 91e–92c). However, the existence, number and nature of the primary bodies and their Ideas are not entailed by the existence of the Idea of animal (or of soul-in-something). For the differentia of the kinds of animals are not, for Plato, differences contained within the general notion animal, such that, if it did not exist, the differentia would not exist. So one of the conditions for parts of Animal being Platonic species fails, namely, the condition that the 'genus' entails its 'species' as number entails 10,000.

Neither, though, does the second condition hold, that of the 'species' making manifest some structural feature of the 'genus'. For the 'differentia' of types of animals, being (the Ideas of) the primary bodies, are not qualifications or determinations of (the Idea of) soul as soul. Rather, I suggest the Idea of animal stands to its parts in the way in which Plato clarifies how the Idea of difference stands to its parts in the *Sophist*:

> It seems to me that the Idea of difference is cut up into bits, like knowledge. Knowledge, like Difference, is one, but each separate part, in being directed toward some particular subject (τὸ ἐπὶ τῳ γιγνόμενον μέρος), has a name of its own; therefore, there are many arts, as they are called, and kinds of knowledge or science. And the same is the case for the parts of Difference, though it is one (*Sophist* 257c7–d5).

[29] Cornford (2) 40, Allen (2) 88n1.

The Idea of animal, like knowledge, I suggest, has a directedness; its parts stand toward (ἐπι) other objects. The things to which the parts of knowledge are directed (say, protons or plants) establish kinds or branches of knowledge (quantum mechanics, botany) *by their being known*. The objects themselves do not, however, qualify or determine knowledge *qua* knowledge. They are not determinations with respect to what it is to be known, in the way, say, we would qualify opinion *qua* opinion, if we divided it into justified and unjustified sorts. So too the Idea of animal, though it is one, has parts which are each directed to something else, namely, the four Primary Bodies. And each part has a name of its own (gods, fish, fowl, flesh). The objects which determine the differences of the parts do so by participating in the Idea of animal, but they do not qualify it with respect to what it is. They each stand to it in the same way, just as the objects of knowledge all stand to knowledge in the same way. Therefore, the objects do not qualify the Idea with respect to what it is. They do not determine or specify it with respect to what it is to be an animal or soul. The parts will all 'have' animality or soul in the same way. The differentia of Plato's Idea of animal are unlike Aristotelian differentia, which are peculiar to the genus which they differentiate and differentiate it with respect to what it is and so are informative about the genus (*Categories* 3, 1b16–17). Footedness, for example, occurs for Aristotle only in the animal genus, and so, even though it is not entailed by the genus, it does qualify animality with respect to animality. For Aristotle, we learn something significant about animals *qua* animals from their differentia. This is not so for Plato. The Ideas of the primary bodies, as being the differentia of living things, *neither* determine divisions *only* within The Idea of animal (for they equally determine parts of Difference) *nor* do they teach us anything about soul or what it is to live. They do not assist us in understanding what soul is any more than protons and plants aid us in understanding what knowledge is. So the parts of the Idea of animal do not make manifest structural features of the Idea of animal. Neither, as we have seen, are the natures of the 'specific' Animals entailed within the nature of the Idea of animal. So the 'specific' Animals are not species in Plato's technical sense; they are mere parts.

To summarize the doctrine of parts: parts that are *merely* parts of another Idea are merely similar to each other and are so not by what each is, but because they each participate in a 'general' Idea which exists over and above them as itself a unity quite independent of their being its parts. The parts of the Idea of animal, then, as mere parts which are not also Platonic species, are similar to each other in that they have animality severally but in common as an attribute, and they have this not by what they are, but because they participate in the Idea of animal.

If this is the arrangement of Ideas in the *Timaeus*, then when the Demiurge looks to various Ideas of animals to seek out an Ideal model for his makings, it is not at all surprising that he should look to the 'general' Idea of animal, rather than to any of the 'specific' or partial Ideas of animals. If he is to make the world an animal or an ensouled thing, he will look straight away to the Idea of animal or soul rather than to any Ideal animal which merely participates in the Idea of animal. For the 'specific' kinds of Animals contribute nothing to our understanding of Animal, since they are Animals not by what they are but by participating in the Idea of animal itself. The opposite would be the case if the model of arrangements of specific and generic concepts were here Aristotelian in structure. The Demiurge on that model would have to look to the species of animals. For on the Aristotelian model, since differentia are peculiar to the genus they qualify, the species which are determined by them are completely exhaustive of the content of genera. Hence, for Aristotle it is more informative for knowing what-animality-is to examine species than genera.

Further, for Plato, though not for Aristotle, it is possible to have an animal that is not any particular kind of animal. For Plato the World-Animal is an instance of the Idea of animal without being a particular kind of animal, that is, without in addition participating in one of the 'specific' Ideas of animals. Since for Plato "animal" simply means "soul-in-something," the World-Animal will simply be soul in whatever there is for it to be in, that is, any and all types of body. And thus we find the World-Soul extended through all the types of matter or elements that there are. The 'specific differences' of Plato's animal 'genus', in being simply specifications of material environments for soul rather than qualifications with respect to soul *qua* soul, are not specifications which are necessarily prior to the instantiation of the genus. Again just the opposite is the case for Aristotle, for whom specific differences of a genus are prior to the genus and indeed constitute the actualization of the genus, so that the genus cannot be instantiated except through one of its species (*Categories* 5, 2a36–2b1).

Since for Plato the 'specific' Animals contribute nothing to the understanding of animality and since animality need not be instantiated indirectly through its 'species', the Demiurge is fully motivated, in searching for a model, to look to the 'general' Idea of animal, that is, to The Standard Animal, and also is capable of making an instance of it alone independently of its parts.

VII. COMPLETENESS

The objection might be raised that Plato's explicit reason for having the Demiurge look to the 'general' Animal rather than to 'specific' Animals is that "no image of an incomplete thing (ἀτελεῖ) is ever good" (30c5).

And so it might be claimed that I have failed in my reckoning of the Demiurge's motivation to give an account of how completeness and incompleteness enter the argument here. Further, commentators find this sentence (30c5) to be decisive evidence for the view that uniqueness in the argument is to be taken as an internal property of the Idea of animal. These commentators (Patterson, Parry, and before them Archer-Hind) get this reading by taking "completeness" to mean "wholeness" and by taking "whole" to mean "composite of essential constitutive parts." This result will follow, naturally enough, if one has misconstrued "parts" of Ideas to be "essential constitutive parts" of Ideas. But this error aside, I think it can be shown that Plato associates completeness (τὸ τέλεον) more closely with its root senses of "having been accomplished," "being perfect," "having become (self-)fulfilled," and "being (self-)sufficient" than with "wholeness," a notion (τὸ ὅλον) which he might easily have introduced here had he wanted to do so. I suggest that the notion of completeness, stripped here, by its position within the Ideal theory, of its associations with projects and processes, means self-sufficient, self-substantial, fundamental, and non-derivative.

When Plato describes the Idea of animal as "complete in every way (κατὰ πάντα τελέῳ, 30d2) and as "the entirely complete animal" (τῷ παντελεῖ ζῴῳ, 31b1), he is referring to the Ideal status of the Idea just as in the *Republic* Plato refers to each Idea canonically as "that which completely is" (τὸ παντελῶς ὄν, *Republic* V, 477a3; τελέως ὄν, *Republic* X, 597a5). There is nothing in the context of these passages of the *Republic* to suggest that either the ontological status of the Ideas or the nature of each Idea is determined by the Ideas somehow being wholes of some sort of parts. Indeed Plato's substitution in the *Timaeus* of "non-partible being" (τῆς ἀμερίστου οὐσίας, 35a1–2; cf. a5, 37a6) for these canonical expressions of the *Republic* seems to rule out that Plato is thinking of completeness in terms of wholes-of-parts. What though Plato positively means by these, and allied canonical expressions (τὸ εἰλικρινῶς ὄν, *Republic* 477a7, 478d6–7, 479d5; κλίνης ὄντως οὔσης, 597d2) is much debated. But I suggest that Plato gives us an explicit and sufficient clue to what he means when saying that something "completely is." In his remarks introducing the Form of the good in the *Republic*, Plato makes the following generally overlooked claim:

> Any measure (μέτρον) which falls short at all of being (ἀπολεῖπον τοῦ ὄντος) does not exist at all in the way measures do. *For* nothing which is incomplete (ἀτελές) is the measure of anything (*Republic* VI, 504c1–3).

In the *Philebus* we also find completeness appearing as a mark of standards or measures (20d1–6, cf. 65b8; 60c2–4; 66b1–2). In these passages completeness is strongly allied to sufficiency (τὸ ἱκανόν) and is not treated in terms of wholeness.

The context of the *Republic* passage (504c) makes it difficult to suppose that "being" is used there in an incomplete sense, that is, where "to be" is a copula with a variable for which all possible predicates may be values as its unstated complement. And it is not clear what such a reading would possibly mean. For, on the one hand, if we take "to be" here to mean just "to be F" without any further qualification, then the sentence is simply saying that a thing which lacks some property (to be specified) will not be a standard. But simply having a certain property is not a mark of a standard or measure. If this were the case, everything would count as a standard and these sentences would defeat themselves. For they are supposed to be distinguishing what is from what is not a standard. On the other hand, if, along the line of Vlastos and his followers,[30] we take the Platonic "to be" as elliptical for "to be completely F" and construe this to mean "to be genuinely F," "to be F *par excellence*," or "to be F to the greatest degree possible," such that the measure or standard for F-ness is that which is F to the fullest *degree*, then the sentences become circular. For the sentences elaborate completeness in terms of the expression "to be." On this reading, the sentences say that that which is a complete F is that which does not fall short of being a complete F. This reading renders the sentences trivial.

Nor does it seem that "to be" here is to be taken in some complete veridical sense, as meaning *either* "to be true" *or* variously "to be that which makes statements/thoughts true," "to be the case," or "to be as it is (said to be)." For we do not find in the context verbs of thinking or saying in search of conditions of explanation for the truth values of the statements which they assert. And moreover though measures and standards play a crucial role in the establishment of truth claims for Plato, one standard by itself does not make a state of affairs, that is, something which may be captured in a propositional form. A standard or measure simply by itself *does* fall short of being that which makes statements and thoughts true. So the use of "to be" here does not seem to have a complete veridical sense.

Rather, "to be" seems to be used here in a complete, existential sense.[31] The sentences are saying that complete reality belongs to measures or standards *qua* standards. So "completeness" means something like "self-sufficient," "fundamental," "basic," or "non-derivative." Plato, then, in saying that standards completely exist, is saying that standards or measures are for him the fundamental entities among the furniture of his universe. By extension, he is saying that all other things as falling short

[30] Vlastos (5)(6), Ketchum, Santas, Annas (2) 198–203.

[31] For an elaboration of the complete, existential sense of "to be" here and an independent defense of the Platonic "to be" as a complete existential sense, see section IXBiii.

of full being are derivative, are dependent for their identification and intelligibility upon standards, and are what they are as the result of some relation they hold to standards.

When then Plato refers to Ideas as "what completely is" or as "entirely complete being," he is assigning to them the role of being the fundamental entities of his metaphysics and they have this role as standards or measures. When Plato specifically refers to the Idea of animal as that which is entirely complete, that which is complete in every way, and that which has impartible being, he is referring to it as a standard or measure. The 'specific' Animals are incomplete in that they merely participate as multiple parts in the Idea of animal. They, therefore, are not themselves adequate as standards for animality. It is not because they are constitutive of the Idea of animal that they are parts, and so are incomplete. It is their status as mere participants that makes them incomplete, derivative, and plural.

To say then that the Demiurge in his craftings looks to the Animal that is complete in every way is just to say again that he looks to The Standard Animal. It is *qua* The Standard Animal that the Idea of animal is sought out by the Demiurge for his unique creation.

VIII. WORLD FORMATION

As we have seen, uniqueness is a standard-establishing property of the Ideas, which is entailed by their status as standards. Uniqueness is also a standard-enhancing property for those standards which are mired in the vagaries of the corporeal. It is far better for both our cognitive and practical projects to have immanent standards or measures be unique rather than to have many standards for each type of thing measured. If our standards of time are, say, the many watches in a pawnshop, we will be hard pressed to know what time it is, how long we have been working, when to meet our friends, whether we are late for the opera, and the like. If our measures of distance are, say, city blocks, we will have similar problems. Given, then, the desirability for immanent standards to be unique, the only question that remains is whether such an achievement is possible, when proper acknowledgments are paid to the way the world is.

In the case of temporal measures, the Demiurge cannot introduce uniqueness into the phenomenal realm, given that Plato had to save the appearances with which that realm actually presents us and that it indeed presents us with a variety of celestial 'clocks', indeed at least eight of them (38d, 39b, c). The best Plato can do is to 'incorporate' the plurality of clocks into his teleology by assigning its source to the cussedness of the corporeal nature of the phenomenal realm, with which the Demiurge must work. And so in the celestial realm we are left with a situation rather like the pawnshop with its many watches running both severally (38d4–6) and collectively at different rates (39c6–7). The same claims are made of the various temporal paradigms in the *Republic*, where their deviations from conditions of constancy are directly attributed to their corporeality (530a7–b3). And the result (as we would expect on my reading) is a reduced intelligibility available for those who try to tell time accurately: the measures of time are mutually incommensurable (*Republic* 530a7–b3, cf. 531a1–3) and are bewildering due to their number and intricacy (*Timaeus* 39d1–2).

In the case of the geometrized primary bodies (earth, air, fire, water) as well, the Demiurge could not make a single instance of each. Primary bodies are images of forms that supervene directly on the receptacle of space and fill space to the greatest degree possible (51a2, 52e2, 58a7, 80c3). So if the Demiurge were to try to arrange things such that there were just one very large instance of each type of geometrized primary body, each a Platonic solid, other smaller instances would immediately appear in the interstices that, by the logical necessities entailed by the geometry of the Platonic solids, would exist among the giant geometrized particles.

However, in the case of the World-Animal as an immanent standard, the nature of the corporeal does not present an impediment to the Demiurge's designs. The World-Animal is essentially a psychic entity and only incidentally a material one. The incursions of the bodily into the World-Soul have the effect of dulling and disrupting the rationality with which the Demiurge invests it (at least in the *Statesman*, 273a–e), but these buffetings of the World-Soul by the corporeal do not tend to split up the World-Soul into a plurality. The material conditions of the world do not prevent the Demiurge from making the world a unique animal, if he has a reason to do so.

Further the need for Plato's cosmology to save the appearances with which the phenomenal world presents us is no bar to the Demiurge making the world into just one animal. It is true that it is highly counter-intuitive (for the Greeks as well as for us) to think of the world as one animal. We believe that most of what there is around us is quite inert. But it is not primarily our beliefs that Plato's reformist doctrines must save; it is the appearances. We would rightfully reject Plato's physical theories if, based on them, there could only be, say, two planets or two elements or two dogs. For we actually see more than two of each of these. But the same simple defeat by the senses will not be possible in the case of Plato's theory of soul or of his theory of sub-atomic particles (54b–55c), since on these theories souls and sub-atomic particles are not direct objects of perception. There is nothing in our sense experience which decisively prevents us from believing that the world is one animal. And this is all Plato needs. In general, then, the conditions of production in the world are such that they do not prevent the production of the world as one animal.

Moreover, the logical attributes of the concept 'animal' for Plato also offer no such prohibition. For Plato, 'animal' is not a concept such that if it is instantiated at all, it must be multiply instantiated. Some concepts are of this sort (e.g., end-of-bridge, pole of a polar opposite), such that if there is one of the kind, there must be at least one other. Now 'animal' under usual circumstances is generally thought to be a concept of this sort. I say under usual circumstances, for one can imagine the unusual circumstance in which one of the giraffes on Noah's ark dies and leaves but one giraffe. But conditions of gradual extinction aside, given the reproductive nature of animals, if you have any of a kind, then you have more than one of that kind. For Plato, however, reproductive capacity is not an essential attribute of animals (see 90e–91a). Indeed at 42a–c, Plato makes one's gender derive from one's morals, rather than from biophysical considerations. Plato leaves the World-Animal utterly sexless (33b–34a), at least in part perhaps to get around the sticky metaphysical difficulty with which our ordinary conception of gendered animals would confront his uniqueness argument.

There are, then, neither material nor formal constraints upon the Demiurge which prevent him from making a unique instance of Animal, if he has a reason to do so. His reason to do so, as we have seen, is to increase the intelligibility of the phenomenal world by introducing standards into it. And so he makes the world like the Idea of animal *qua* standard. And he does this by making but one instance of the Idea, since uniqueness is a formal property the presence of which makes standards serve better as standards. He, therefore, makes the World-Animal-in-general one of a kind.

It should be noted that this line of argument only gives a reason why the World-Animal is unique. It is not an argument which explains why the World-Animal is stretched throughout and encompasses the whole of the corporeal realm (*Timaeus* 34b3–4, 36e2–3). The uniqueness argument would be compatible with the allocation of the single instance of animal-in-general to just some portion of the corporeal realm. I discuss the reason for the all-encompassing nature of the World-Animal later (chapter 9). Basically the reason is that Plato holds one of the primary and peculiar functions of rational soul to be its ability to maintain order against an inherent actual tendency of the corporeal to be chaotic. Therefore, since Plato wishes the phenomenal world to be stable and orderly, he must have the whole permeated with soul. This goal of maintained order and stability could be achieved by having many "contiguous" rational souls blanketing the whole of the otherwise chaotic, phenomenal realm. But the economy-minded Demiurge conflates this end with his standard-establishing project by making the World-Soul *both* unique *and* all-encompassing. Since souls for Plato do not, as they do for Aristotle, have a proper matter, the World-Soul need not be attached to any particular type or portion of the world's matter, so there is nothing which prevents the World-Soul from encompassing all the matter there is. That it encompasses all the matter there is, though, is not to be construed as the reason for the world's uniqueness. If this were the case, the presence of the World-Soul would be quite gratuitous, since, as we have seen, the world is said to contain all the matter that there is quite independently of the presence of the World-Soul and the reasoning of the unique world argument (32c–33a).

It is not the case for living things as we know them that they could ever be all encompassing in the way the World-Animal is or even be much larger than they in fact are. The nature of bodily functions of living creatures, in particular respiration, together with the mathematical properties of physical existence prevent living things from being very large. The reason for this is that the need for respiration is a function roughly of weight and derivatively of size, the increase of which is charted by a cubic function, whereas respiration itself is a rough function of surface area, the increase of which is only a quadratic function. So, even with adaptations like leaves and involuted lungs, creatures as they increase in size have a need for respiration which quickly overtakes their ability to meet that need. There are, therefore, never going to be any flora and fauna the size of Mt. Everest, let alone the size of the universe. Only a creature like Plato's World-Animal which is not defined by bodily functions, and indeed is largely devoid of them, could be all-encompassing, and so fulfill the regulatory function which Plato wishes to assign to (rational) soul.

It is this regulatory function of the World-Soul which explains its relation to the planets. The circuits of the World-Soul are made to mesh with and integrate with those of the planets in order to maintain such stability and order as the wandering planets have, so that the planets may in turn serve as the standard measures of time. It is not the feature of the planets as also eventually (though incidentally) being portrayed as ensouled (38e5–6, 40b–d) that is relevant here (36c–d, 38c–d). The World-Soul is not being viewed as a system of the souls of the planets or indeed of any other souls. Rather the World-Soul, in its aspect of being all-encompassing, explains the general orderliness and regularity of the material realm, which, if left to itself, would tend to run off to excesses. We see, then, that two of the oddest features of the Platonic cosmology—that there is an animal that is unique and that this animal encompasses the whole universe—are fully motivated within Plato's larger metaphysical concerns.

IX. SOME PLATONIC TENETS

A. Divinity and Cognition

The Demiurge is introduced in the *Timaeus* first by an enquiry into the nature of his model. It is claimed that there are two kinds of models (παραδείγματα). One kind is eternal; these models are the Ideas. And one kind is generated (γεγονός) (28c6–29a2). It is claimed that since the world which the Demiurge builds is to be good, clearly he used the eternal model, and in any case it is courting blasphemy (μηδὲ θέμις) for anyone to say that he used a generated model (29a2–4). It has been suggested that the presence here of generated models is a "patent fiction" hypothesized solely for the sake of argument.[32] On this account, the rejection of generated models as possible objects of the Demiurge's craftings is construed merely as a rhetorical flourish aimed at heightening the awareness that all paradigms are non-sensible and eternal. This hardly seems the likely sense of the passage, given the central position which (demiurgically) generated, sensible paradigms hold in the epistemology of the *Republic* (529d7–8, 530a6, 530b1–3), the metaphysics of which is reproduced here in the opening pages of Timaeus' discourse (27d ff). The reason for the rejection of the generated paradigms as models for the Demiurge is not that they are a patent fiction, but that given an option between the two kinds of models, the Demiurge would certainly avoid the paradigms that are enmeshed in the vagaries of the material.

Nevertheless, the presence of generated, sensible paradigms (29a2) here at the start of the *Timaeus* is paradoxical. For if "generated" means "generated by the Demiurge," as our analysis of the *Republic, Philebus,* and the rest of the *Timaeus* might suggest, then generated paradigms could hardly be candidates as *models* for his crafting activity, since they are *products* of the same activity.

One might try to resolve this paradox by following the overwhelming tendency in Platonic scholarship to take the figure of the Demiurge non-literally, that is, as not being intended as an actual component of Plato's cosmological commitments. There has been a dazzling array of readings of Plato's cosmology which take Plato as *meaning* something quite different than what he *says* when he describes the Demiurge. The Demiurge has been taken as a mere doublet of the World-Soul (Archer-Hind), or for the rational part of it (Cornford), as a general symbol for any craftsman-like activity (Cherniss, Johansen), as only a hypothetical entity serving merely as a literary foil in the exposition of the human statesman and the

---

[32] Robinson (1) 67.

World-Soul (Herter), and as a "sublation" of the World-Soul (Rosen).[33] These diverse strategies should be seen, I think, largely as (unneeded) charitable attempts to distance Plato's thought from Christian thought and more generally as attempts to reduce the number of unfashionable theological commitments in Plato's cosmology. If my reading of the activity of the Demiurge is correct and he is not essentially just a cosmetologist or interior decorator out to make things ravishing, but is primarily a maker of immanent standards, which enable our having true opinions, then much of Plato's epistemology would also seem at first blush to hang on eccentricities of his theology. So a non-literal reading of the Demiurge might appear attractive. However, given the frequency and prominence with which the Demiurge appears in both mythical (*Timaeus, Statesman*) and non-mythical settings (*Republic, Sophist, Philebus*), his excision from the Platonic metaphysics through what invariably turn out to be idiosyncratic hermeneutics seems a desperate measure.

There is another way out of our paradox of the generated paradigms, a way which would spare Plato's epistemological claims from depending on quirks of theology. When Plato speaks of generated paradigms in our passage he may simply mean that they are part of the world of becoming, the comings-into-being and perishings of the phenomenal realm (28a3).[34] And by allowing that the Demiurge might take these as his paradigms, Plato may very well be suggesting that if our minds are somewhat like the Demiurge's, then it is within our creative, investigative capacities to *discover* standard cases within the phenomenal realm and that immanent standards need not only be viewed as *products* and immediately provided gifts of another's construction.[35] I wish tentatively to suggest that something like this is in the offing in the Platonic texts.

In at least three diverse passages, Plato claims that the human mind

---

[33] Archer-Hind 37–40; Cornford (2) 34–39, 76; Cherniss (1) 607, cf. 425; Herter (2) 106–17; Johansen 86; Rosen 75–76. For refreshing exceptions to this non-literalist trend see Robinson (1) chapters 4, 8, 11 and Taylor (1), (2). It should be remembered, however, that Taylor does not think that the *Timaeus* is an account of Plato's own views.

[34] There is a great deal of critical confusion over the various possible senses of "becoming" in the opening pages of Timaeus' discourse. This, I think, is due to the text itself being confused. But in any case, not all the references to "becoming" can mean "is being produced by demiurgic activity" as is shown by the mention of γένεσις at 52d3 to describe the world "even before the heaven came into being." For the contrary position though, see Robinson (3) and cf. Johansen 71.

[35] Letwin has made a suggestion along these general lines, claiming that the human mind, which he identifies with the fifth 'class' in the *Philebus,* is the capacity to analyze what the demiurgic mind synthesizes in its constructions. Letwin 187–206. Letwin, however, produces no evidence to support his contention

has a sort of dim foreknowledge of concepts, a foreknowledge which, on the one hand, is possessed independently of also possessing whatever full Platonic knowledge is (direct acquaintance with Forms or whatever) and, on the other hand, is not derivative from sense experience. This dim foreknowledge allows one to begin to make somewhat accurate judgments about the phenomenal realm (*Republic* VI, 505e1–3, foreknowledge of good; *Phaedrus* 250a–b, foreknowledge of beauty and other normative notions; *Philebus* 35a–b, foreknowledge of the nature of the objects of desire). Admittedly each of these passages emphasizes the tentativeness of the judgments made by using the dim foreknowledge. The *Republic* passage, for instance, explains that the judgments made of the phenomena using this foreknowledge of good do not constitute fully accurate identifications, since the judgments cannot answer the "what is it?" question (505e2) and do not constitute "firm/certain belief" (πίστει μονίμῳ, 505e2–3). But that we are able to make somewhat accurate judgments at all concerning the phenomenal realm is dependent on our possession of some foreknowledge.

Plato is suggesting, I contend, that the human mind, prior to its interaction with the phenomenal realm, contains something like rough templates by which it, as it scans the phenomenal realm, is able to pick out phenomena which, upon closer approach, turn out to be such as to serve as standard cases for their kind. These standard cases, which are discovered by the aid of dim foreknowledge, in turn enable the mind to make identifications of other phenomena, acting in this way like one who has discovered blueprints carefully "drawn by Daedalus or some other craftsman." The rough templates allow the mind to act like current sophisticated radar systems which by the use of templates or profiles are able to determine not only the location of objects but also can determine whether, say, floating objects ahead which might be any one of several types of things, are indeed going to turn out to be, upon closer approach, a flotilla rather than a herd of whales or a group of icebergs. The rough templates allow one to come into direct contact with the relevant cases of the object of one's inquiry, and then in turn to use these as standard cases for further identifications.[36]

In the *Philebus,* Plato specifically entertains the *human* invention of measures or standards from out of the phenomena. Plato catalogues a

---

that the fifth 'class' is human thought, other than to mention the single reference to the fifth 'class' at *Philebus* 23d9–10, lines which make no mention of human thought and which fail to give any specific sense to 'division'.

[36] In chapter 14, I tentatively suggest that the objects of the third level of the Divided Line, the object of διάνοια (*Republic* VI, 509d–511d), are indeed themselves mental entities, which operate as rough templates or fuzzy concepts in our identifications of phenomenal objects.

whole host of crafts and sciences for which measures are not immediately given in sense experience. These include medicine, agriculture, piloting and generalship, though music is taken as paradigmatic for the whole class (56a–b). These endeavors are said just barely to count as sciences (55e7–56a1); for they cannot appeal directly to a measure (οὐ μέτρῳ, dative of respect, 56a4). However, by sifting and sorting through the things subject to the more and the less and by guesswork based entirely on working over empirical data (55e5–6), these marginal sciences are on their own able to discover the relevant measure (τὸ μέτρον, 56a5) for their art or craft, though it is not a very clear or reliable one. In the *Timaeus*, some measures and proportions are said to exist by chance even in the pre-cosmos (69b6). In the *Philebus*, the inexact sciences almost by chance bump up against these measures that themselves exist by chance. But even the measures and standards used by the architect and builder, whose crafts are more scientific than music and the others, appear to be the discoveries or products of human ingenuity—the straight-edge, peg-and-cord, compass, plummet, and set-square (56b–c).

If this sketch of the operations of the mind on the phenomenal realm is correct, then the theology of the *Timaeus* could be sheared off leaving intact the doctrine of immanent measures and standards as the objects that enable human's cognition. The sketch admittedly makes Plato, as far as cognition of the phenomenal realm is concerned, a proto-Kantian or proto-structuralist to a degree greater than any current critical tradition would be willing to accept. The sketch differs from a strict Kantian view in that the dimly foreknown concepts are not (restricted to) organizational and categorical notions (substance, quality, causation, etc.), but explicitly include some of the content of what is discovered, in the way that templates do. If the Good in the central books of the *Republic* includes within it the constellation of metaphysical properties which make up the Ideality of the Ideas (as I think it does),[37] then our foreknowledge of good (*Republic* 505e) will also provide us with foreknowledge of the standard-establishing properties, which will be requisite (along with the foreknowledge of the content of concepts) for discovering or inventing standard cases.

Further, I do not wish to suggest that past Kantian interpreters of Plato (Paul Natorp of the Marburg School and J. A. Stewart in England) were correct in viewing the Platonic Ideas themselves as nothing more than rules in the mind. Indeed if pressed, I would argue that for Plato we each have the ability to find standard cases in the phenomenal realm just exactly because the Ideas exist, the same for all, independently of the human mind; but, this argument would require a certain construing of

[37] For the Form of the good so construed, see Santas.

Plato's doctrine of recollection, one which I tentatively float in chapter fourteen.

B. Ontology

i. *Forms as Individuals*

I have suggested that Forms taken each as a unique standard are for Plato the fundamental individuals of his metaphysical system. By this I mean that all things other than Forms are dependent for their identification and intelligibility upon Forms taken as standards. I now wish to suggest in addition that the unique world argument shows that Forms as standards are to be taken *fundamentally* as *individuals* rather than as things qualified, so that Forms are both fundamental individuals and fundamentally individuals. This construction will help clarify and make coherent much of the Platonic ontology.

I have argued that Forms as standards cannot be defined as solutions to, or instances of, sets of defining conditions (section IV). I now wish to generalize this claim, by suggesting that Forms are numerically one but not as being *instances* of *any* property and indeed are not even to be thought of as instances of or as possessing the property of which they make possible the identification in other things.

Imagine the following scenario. Someone introduces Romulus to me, alleging that Romulus is an only child (μονογενής). Now if Remus is standing nearby, birth certificate in hand, I would be in a good position to say that the introducer was at least mistaken. If it further turned out that the introducer was fully familiar with Remus, say, by being his parent, then I additionally could reasonably claim that the introducer either was crazy or was lying. If further the introducer, in self-defense, claims that Romulus is an only child just *exactly because* he is an identical twin to Remus, I would have to conclude that the introducer is not crazy, but is intentionally being perverse, hoping perhaps for a chuckle on my part.

Those critics who suppose that Forms have the properties of which they enable the recognition in other things, are committed, it seems to me, to viewing Plato in the unique world argument as taking upon himself the same role as the introducer in our Roman scenario. For if the Idea of animal is such that when it is copied, a similarity obtains between it and the world with respect to animality, and if both the Idea and the world are claimed to be unique each as being the sole possessor of animality-in-general and if this is claimed to be so as a direct result of the similarity of Idea and world, then Plato must be making a bad joke. For one cannot without contradiction claim that two formally identical things are each severally unique with respect to the very property they have in common, and further one cannot in all seriousness draw attention to the contradiction by claiming the two things are unique *because* they are formally identical. Since the unique world argument and its surrounding pages are deadly serious, we need to find some way out of our Roman paradox.

One way out, which will not work, I think, is to claim that the Idea and its instance, though similar with respect to the very property of which allegedly they each are the only possessor, possess it each in a different manner or in relation to some further distinguishing feature, such that each is unique in possessing the common property in some way the other does not. It is in this way that we say the Moon is unique. It is not unique *qua* moon, but in its relation to the Earth it becomes unique. It is the Earth's only moon. On this reading the Idea of animal will be the only animal-in-general which is an Idea and the world will be the only animal-in-general which is a phenomenal object. This way out, however, marks a retreat from the explicit claim that the world is unique *because* it is like the Idea (31a8–b1), and in any case if I am right, Plato intends the Idea and the world to be unique for precisely the same reason, that is, so that they may each be a standard.

The correct way out of the paradox is, I suggest, to recognize that Plato is using "unique" in two subtly but importantly distinct senses. The world is unique in the sense of being one in number and the only instance or possessor *of* its kind. Thus Plato calls the world μονογενής (*Timaeus* 31b3, 92c9), which here has its root sense "only begotten."[38]

Plato, however, crafts the unique world argument carefully so that this term is not used to qualify the model. He uses rather the coinage μόνωσις to describe the uniqueness of the model (31b1). In part the choice is perhaps governed by a desire to avoid associations of generation which attach to μονογενής. In part the choice is perhaps an attempt to signal a slight change of sense. However, since μόνωσις is a nonce word in Plato, its sense, which eventually comes to mean "singularity in virtue of isolation or remoteness," has to be gleaned from the argumentative context.

When that is scrutinized carefully, we find that in ascribing uniqueness to the Ideas, Plato does not mean that each Idea is one in number and the only instance or possessor of its kind, but rather means that each Idea is one in number and the only one *for* its kind, the kind it allows us to identify. For each kind there is but one thing which makes possible the determination that the things which possess the kind indeed do possess it (cf. *Republic* X, 596a6–b4).

Within the unique world argument the subargument showing that the Idea of animal is unique indeed establishes it as unique in the sense of

---

[38] The term μονογενής, which occurs four times in Plato, can also mean "of a single lineage" (so *Laws* III, 691e1) and more abstractly "of one kind." Contextual considerations show that the term does not have this sense at *Timaeus* 31b3, 92c9, nor *Critias* 113d2.

being one-*for*-a-kind and not in the sense one-*of*-a-kind (31a4–7; cf. *Republic* 597c). Important results follow from this. For it means that the world can be unique in the sense of being one-of-a-kind, which is the only way in which an instance of a Form could be unique *qua* instance, and the world can be unique in this sense without, as it were, competition, since the Idea of animal is no longer unique in this sense. But if the world alone is to have the status of being unique in the sense of being one of the kind it is (i.e., of Animal), then the Idea of animal will necessarily not be of that kind. The Idea of animal, therefore, must be fundamentally an individual since it is numerically one, and yet is independent of being of a kind with respect to what it is. Though some things may be said of it (e.g., its formal properties), its "essence" is not an attribute or quality.

Insofar as the uniqueness of the Idea of animal is a formal property of the Idea, what the unique world argument as a whole adds to Plato's arguments for the uniqueness of Forms (*Timaeus* 31a, *Republic* 597c) is a commitment (sometimes read even as implied in those arguments) that each Form is not of the kind of which it allows the determination in other things. Forms are fundamentally individuals.

Plato then resolves our Roman paradox by using 'unique' in subtly ambiguous ways when he ascribes it as a standard-establishing property variously both to Idea and instance. Each Idea is fundamentally an individual and is unique in the sense of being one for a kind; whereas the world is unique as being one of a kind, the only instance of its Form. This means, though, that Plato is neither crazy nor disingenuous in asserting that the world is similar to its Idea and yet that both world and Idea are unique.

If Plato indeed construes Forms fundamentally as individuals, new problems may appear to crop up where old ones were resolved. For the implications of this view for his logic and epistemology will have a tendency perhaps to boggle the mind, at least initially.

If we now ask, for example, in what way Forms for different kinds differ from each other, it turns out that we have no ordinary vocabulary in which to state the answer accurately. They are numerically distinct to be sure. But if what I have said about Forms so far is true, it is now logically impossible to say that they are formally different. For we have eliminated their having properties, with respect to what each is, by means of which they may be formally distinguished. And yet they cannot simply be bare particulars capable of being interchanged without subsequent effect. For they must be discernibly distinct in order to be each for its kind alone. We have to resort to some sort of phrasing like: each has a content which will be distinct from the content of every other Form. Such content constitutes the principle of individuation for Forms. It is that to which we would appeal to explain why a Form is numerically one. Our apprehending that a Form is one derives immediately from our ability to apprehend it as the Form it is.

The names of Forms then will not be disguised definite descriptions. For Forms are not what they are as the result of even partially being solutions to sets of conditions which might be described. If the names of Forms are names in any modern sense, they are most like Russellian logically proper names which pick out individuals as individuals from a field of immediate acquaintance. It will be the content of each Idea showing forth itself as it is which constitutes the distinctness necessary for their being picked out by "this" and "that." Further, though names of Forms do not harbor descriptive elements, nevertheless since there is a peculiar content for each Idea, there will be a correctness associated with the names of Ideas. If we call one Form by a name, it will be incorrect to call any other Form by that name.

It is frequently objected that if Forms or standards are fundamentally individuals and not primarily fulfillments of sets of conditions or are not in some other way subjects of significant descriptions, but are simply given individuals which at best can merely be named rather than described, then they lose their explanatory power and so too their very reason for having been hypothesized by Plato in the first place, and so allegedly fail in their metaphysical and epistemological mission.[39] The answer to this charge is that the explanatory power of Forms or standards lies in their *relations* to other things. Standards allow us to describe and identify other things, and insofar as standards form necessarily related clusters, they can explain causal relations among other things. In this regard, Plato is no more silly than Aristotle or any philosopher who wishes to claim that some principles of explanation must themselves be beyond explanation.

Further it might be argued that if Forms are fundamentally individuals and not essentially things qualified, then, even with possible problems of causal inertness set aside, Platonic Forms will fail to fall within the category of the cognizable on pretty much any theory of cognition which one might pick.

Since Aristotle's day, it has been hard to imagine that anything could be or could be perceived as being one without it also being one of some kind (*Physics* II, *De Anima* II, 12). We can however get some intuitive grasp of what it is like to take in something as being one without having also to consider it as one of some kind, a kind which is capable in theory of multiple instantiations, if we recognize that something like this constitutes our understanding of the way in which we suppose we non-reflex-

---

[39] This charge of explanatory vacuousness has been leveled by Moravcsik. Moravcsik (1) 18–20.

ively recognize individual people as being unique. Our immediate taking in of others is as their being each numerically one but not of some kind. Even if we are pressed into cataloguing a person's accidental characteristics and quirks which are sufficiently diverse to apply collectively only to this person (gait, eye-color, gender, general location, sense of humor, pretensions, etc.), we are not moved to call this compound predicable a kind, even if the catalogued characteristics were jointly capable in theory of duplication. Our tendency is to suppose that even Romulus and Remus are perceived immediately as unique even without appealing to their spatial distinctness or to the order of their births. The reason for our first thinking of people as individuals, rather than as collections of properties, are complex, resting I would guess at the intersection of theology, biology, sociology and ethics. My point is only that we in fact do take in people fundamentally as individuals.

Perceiving or grasping Forms will, I suggest, be roughly analogous to the way we take in people as individuals. Platonic knowledge as a kind of seeing or apprehending with the mind's eye (*Republic* VI, 508d4, *Sophist* 254a10) will be strongly disanalogous, then, to Aristotelian perception or perception-like passive thought (*De Anima* II, 12; III, 4) wherein the cognizer becomes formally identical to the object of cognition. A Form just in being the Form it is has no qualitative nature with which to stand in a relation of formal identity to a perceiver. For it is a 'this' (τοῦτο) with no 'such' (τοιοῦτον), whereas the phenomena as images are 'such-es' but not 'this-es' (*Timaeus* 49d–50c, on which see chapters 3 and 4).

ii. The Third Man Argument

That the Ideas are fundamentally individuals rather than things qualified spares Plato's two-tiered ontology from being committed to the vicious logical regress of the Third Man Argument (TMA).[40] The unique world argument helps us pinpoint where Plato supposes the TMA goes awry when directed at his theory of Ideas. The TMA assumes that a) any Form along with its instances can be taken as members of a set of which all the members severally but in common possess the attribute which makes the instances instances of the Form. The TMA further assumes that b) since in accordance with Platonic principles all attributions of properties to things are made by reference to some Form beyond the set of things which consists of members with a common property, there must be another Form over and above the first. Given a) and b), and if in addition c) the new Form too is formally identical to the members of the earlier set, then there will be an infinite regress of Forms. The regress will be vicious because, in virtue of b), prior members of the regress presup-

---

[40] For texts, see *Parmenides* 132a1–b2, 132d1–133a6; Aristotle, *On the Forms*, ap. Alexander, in *Met.* (Hayduck), 84.21–85.11.

pose (for their identifications) posterior members of the sequence, of which there is no last member. Plato, I suggest, would reject a) and *a fortiori* also reject c). He accepts b). In the vocabulary of the recent critical tradition, a) and c) presuppose self-predication of Forms, that is, they assume that each Form possesses the very property it allows us to identify in other things. And b) presupposes the non-identity of Form and instance, that is, it presupposes that a thing which possesses an attribute cannot be numerically identical with the Form by which we claim the thing has the attribute it has.

Those who suppose that Plato is committed to the TMA, in order to get the requisite premise a) for the argument, must assume that in the unique world argument Plato is reproducing what I called the Roman paradox (section i above). They must claim that the Idea of animal and the world severally but in common possess the attribute 'animal', so that Plato is being intentionally perverse in calling each unique *because* the two are so similar.

If I am right in suggesting that Forms are fundamentally individuals and are not things which possess characteristics with respect to what they each distinctively are, Plato is clearly not committed to and indeed would deny premise a) and its self-predication assumption. For we will not be able to make a mental review of a Form and its instances in such a way that it turns out that we discover them to form a set the members of which each possess some formal identity with every other member. On my analysis, it is just exactly the particular mode which the non-identity of Form and instance takes for Plato that makes Forms non-self-predicating. They are non-self-predicating, because they are unique in the way standards which are not subject to merely relative measurement and comparison are unique. They are one for a kind, but not of any kind with respect to what each alone is. The interpretative error most widespread in the literature which views Plato as committed to the TMA is the confusion of perfect instances of Forms with Forms themselves, that is, the confusion of what I have called an immanent measure or standard case (a thing which possesses a property to the fullest extent possible) with that non-derivative standard by reference to which we say that an immanent standard possesses a property fully. If we suppose that Plato thought that the sorts of things we can reasonably say of The Standard Pound are no different from what we can say of the pound measure which you and I might use, then he surely will be committed to the confusions of logical types we find in the TMA.

iii. Real Being

If Forms are fundamentally individuals, then we will again see that the Platonic sense of "to be" is a complete existential one, and we will be able to give a fuller account of this sense and of what Plato means when he says that each Form "really is" (κλίνης ὄντως οὔσης, *Republic* 597d2; τὸ ὂν ὄντως, *Philebus* 59d4; οὐσία ὄντως οὖσα, *Phaedrus* 247c7).

In the extensive debates over the sense of the Platonic "to be," the range of possible senses of "the Form F *is*" has become saturated: every possible sense of the Greek "to be" has been ascribed to the Platonic "to be." The possible senses of εἶναι are:

1. incomplete copula:	"to be" = "to be F" (taken as the Platonic "to be" by Vlastos *et al.*).[41]
2a. complete first-order veridical:	"to be" = "to be so," "to be as it is said to be," "to be the case" (i.e., "is the state of affairs which true propositions describe").[42]
2b. complete second-order veridical:	"to be" = "to be true" (as applied to propositions).[43]
3. complete existential:	"to be" = "to exist" (old guard unitarian critics).[44]

If Forms are standards or, more especially, are fundamentally individuals, the Platonic "to be" will have to be a complete existential sense. The old guard unitarians, I think, are right on this matter. The Platonic "to be" is *some* sense of "to exist."

If Forms are not fundamentally (if at all) things qualified with respect to what each particularly is, then the Platonic "to be" applied to the Form of F cannot mean "to be F" and so *a fortiori* the distinctive way in which a Form is said to be, i.e., 'completely' or 'really' cannot mean "to be F *par excellence*," or "to be F to the greatest degree possible" [therefore, not-1].

Further, if Forms are fundamentally individuals, then a Form by itself, though it is said "to be," does not constitute a state of affairs which can be captured in propositional form. So it does not seem that the Platonic "to be" is a first-order veridical sense ("to be the case"). Only a network or combination of Forms could be said to be in this sense. And it is pretty clear that the Ideas as a whole are said to be because each individual Idea is said to be and not vice versa. In general the view that the Greek "to be" is a first-order veridical sense fails to give full weight, or indeed any weight, to the role that unity plays in defining 'being' at least in Parmenides and Plato[45] [so, not-2a].

---

[41] Vlastos (5)(6), Ketchum, Santas, Annas (2) 198–203.

[42] Kahn (1), (2).

[43] Fine (1).

[44] Frequently only implicitly assumed, but see for example, Cherniss (4) 130–32, Allen (3) 51–52 and especially 57–58.

[45] See Kahn (1).

When it is claimed that the Platonic "to be" is a second-order veridical sense which, when taken together with the qualifiers "really" and "completely," is solely applicable to the Forms, what is meant is that *all* propositions which actually have a Form as their subject's referent will invariably be true (whereas *some* statements about a phenomenal object will be true, others false). But as with sense 2a, a Form just by itself, though it *is*, does not establish the truth of even a single proposition (regarding what it alone is) let alone establish the truth of a whole field of propositions [therefore, not-2b].

Therefore, Forms must be said "to be" in some complete existential sense. The Platonic "to be" is some sense of "to exist." However, if I am right about Forms as individuals, the sense of "exist" here cannot be our modern sense of "to be" where "to be" means "to be an instance of a concept" or "to be the value of a variable.[46] If Forms are fundamentally individuals with respect to what each one is, then they are not even candidates for serving as things over which we may quantify. On the modern account of existence, the instances of Forms will exist, but Forms themselves will not exist. With respect to being the Form it is, no Form possesses properties which can be cast as predicates in such a way that the Form's name may be said to provide a value for their variables. Further, however one construes those among the Great Kinds in the *Sophist* which are said of all forms (namely, Being, Sameness, Difference, and Rest), it is clear that Plato supposes Forms to exist because they participate in Being rather than because they can be values for bound variables of the predicates 'same', 'different', and 'at rest' (*Sophist* 252a, 254d, 256a, e). Ironically, on my account Plato turns out not to be a Platonist, as "Platonist" is used in current discussions of number theory, wherein to be a Platonist is to be committed to quantifying over abstract entities.

I suggest that the Platonic "to be," because it applies directly to individuals as individuals, must mean something within the constellation of notions "to be actual," "to be there in such a way as to provide an object to point at," "to present itself." When the adverbial qualifications "really," "completely," or "purely" are attached to this sense of "is," the compounded designation will mean "is this way (actual, self-presenting) *on its own* or *in virtue of itself*," or "is there to be picked out independently of its relations to anything else." These adverbial qualifications are basically equivalent to the Platonic καθ' αὑτό, especially when it is contrasted with πρός τι.

The main engine for the view that the Platonic "to be" is an incomplete predicative sense is that an existential sense would be incompatible with Plato's various claims that different things may have different *degrees* of

[46] For a denial that the modern sense of "to be" is the sense used by Greek philosophers generally, see Kahn (1) 323–25.

being or admit of *more or less* being (e.g., Republic V, 479c–d1; VII, 515d1–3). This is the view of Vlastos and especially Mourelatos.[47] Only the predicative sense—"to be F"—so it is claimed, can properly capture the notion of different things being in different degrees, since only predicates (or at least some of them) pick out properties which can be manifested in degrees. Allegedly existence cannot admit of degrees, since "existence" is not a predicate. However, this allegation will be true only if "to exist" is construed in a post-Kantian sense as "to be an instance of a concept," "to be the referent of a subject of which predications are made" or more formally "to be the value of a bound variable." And yet, if "to exist" is taken (as I have suggested Plato intends it to be) to mean "to be substantial (independently of its relations to anything else)" or "to be there on its own in such a way as to be pointed at" or the like, then this sense is quite compatible with *an* understanding of different things possessing different degrees of being—an understanding on which the various degrees need not form a *continuous* scale.

The telling example of this understanding for the Platonic metaphysics is the following. A shadow or an image in a mirror or a dream object will *be less* (*real*), in the requisite sense, than its original. And the original will *be fully* or *be completely* (*real*), again in the requisite sense—being actual, substantial, being there self-presentingly on its own. For the original's substantiality is not further dependent upon something else in the way the image is dependent upon it. And yet there is no continuous scale of degrees between the grade of existence of the image and that of its original. This account, then, of the Platonic "to be," which simply appeals to the intuitions that stand behind some of our quite ordinary linguistic conventions concerning "being" and "real," captures better than does Vlastos' the central metaphor of original and image by which Plato chiefly conveys his metaphysics of Forms.

Critics like Vlastos and Mourelatos too hastily assimilate the metaphor of original and image to the predicative sense of "to be." Vlastos supposes that the artifacts that cast shadows in the Cave (*Republic* VI, 515d1–3) "are more" because they have more properties of the thing of which they themselves are images than do the shadows that they cast. So the artifact is more F than the shadow because it shares more properties with the thing in the daylight world that is F. But this is not necessarily true. Suppose our artifact is a ten-inch tall pink and purple polka-dotted puppet of a horse. The shadow of it on the wall may well have the size that real horses have while the puppet does not, and the shadow, unlike the puppet, comes in a color, black, that horses actually come in. So in these respects the shadow is more like a real horse than the puppet. But the puppet is more real in the sense I'm advancing, because it can subsist

47 Vlastos (6) 60–63, Mourelatos 65.

and shine forth without the shadow, but the shadow cannot subsist and shine forth without it. The shadow is ontologically dependent for its ability to shine forth both on the persistence of its original and the presence of a medium on which it may appear.[48]

Plato does not suppose that his talk of a thing admitting more or less of something else entails that the something else consists of a scale of *continuous* gradations (though that will be the usual case). For in the *Statesman*, Plato ranks number (i.e., integers) along with length, breadth and thickness (or swiftness) as examples of things that admit of the more and the less (284e) and yet he is as fully aware as Aristotle that numbers do not admit of continuous variation. Five fingers are not just sort of odd; they just are odd and not even a slight bit even or a slight bit six in number (*Phaedo* 103e–104b, 104d–105b, 105d, 106b–c, and especially *Cratylus* 432a–b). For Plato, being, like integers, may be manifest in non-continuous degrees.

Plato is a Neoplatonist to the extent that he thinks that being is sometimes a predicate; however, he fails to be a Neoplatonist in that he does not suppose that an examination of any two grades of being will always reveal some third intermediary grade.

This reading of the Platonic "to be" accounts nicely for what Plato says about the sense in which things other than Forms *are*. The phenomena exist on this account, but they do not *fully* exist. They are there to be pointed at but not as the result of their status as phenomena. For they are doubly dependent on other things. They depend both upon the receptacle or space, which serves as a medium for their reception as images, and upon the Forms for their ability to shine forth (*Timaeus* 52a–c).

In calling the receptacle itself a 'this', Plato seems to want to assign to it the same full reality which he assigns to the Forms (50a1–2, a7–b2). Plato's confessed trouble with designating clearly the mode of cognition of the receptacle (52b1–2) then arises not because the receptacle totally lacks any proper attributes or content (50b–c, d–e, 51a7), but rather because, though it is a 'this', it indeed does not shine forth or present itself to us. When we look to it, we do not see it—we see what is *in* it (52b3–5). When we point at it, we necessarily do so indirectly.

If I am right in the characterization of the sort of existence which Forms must have as the result of being fundamentally individuals, the mode of cognition of Forms must be a form of unmediated acquaintance, operating, as I have suggested, on a rough analogy with the way in which without a moment's reflection and really without doubt we recognize people as individuals.

---

[48] See Lee (1), on whose analysis I draw here.

TWO

Plato on Time and Eternity*

SYLLABUS

I. INTRODUCTION

Considerations of intrinsic interest aside, Plato's views on time are important in that they are embedded in and, I suggest, can only be coherently interpreted in their general features as a part of Plato's broader metaphysical project and so throw light on that project as providing a highly illustrative instance of its workings.[1] The sense in which time for Plato is jointly a product of a craftsman God and an image of an Idea is paradigmatic for all cases of such production and imaging (*Timaeus* 39e3–4). Moreover, in its details, the discussion of time in the *Timaeus* (37c–38c) is the only place in the corpus where Plato provides any explicit discussion or analysis of the cluster of attributes 'eternal', 'permanent', and 'immutable', which he frequently assigns to the Forms as

---

* This chapter was first published in *The Platonic Cosmology* (1985).

[1] I have occasion to refer to the following literature on Plato's views of time and eternity: Cherniss (1) 212 ff., 417 ff., (5) 234–37, (8) 340–45; Crombie 208; Cornford (2) 97–137; Guthrie 299–301; Hintikka; Kneale 93–94; Owen (2) 317–40, especially 333 ff.; Robinson (1) 95–97; Tarán 378–80&nn; Vlastos (2) 385-90, (3) 409–14; von Leyden; Whittaker; Zeyl (2) xlii–xliv.

properties which they have just as being Forms.[2] These assignations are so frequent and so central to Plato's understanding of Forms that they come to stand on their own as periphrastic designations for the Forms themselves, and so they come close to being technical expressions for Plato.[3] However, not much can be gleaned about the sense of these terms as technical expressions from their usages in describing and designating Forms.

In the *Timaeus* passage itself we see Plato innovating and struggling with a seemingly recalcitrant vocabulary. He uses an adjectival coinage αἰώνιος ("eon-ly") of both the Ideas and time (37d3, d7; only other occurrences in Plato, *Republic* II, 363d2, *Laws* X, 904a9); he further coins an intensified form of this adjective διαιώνιος ("very eon-ly"), using it only in this passage and only of the Ideas (*Timaeus* 38b8, 39e2). Little if anything, though, can be made of such linguistic tinkering. More important perhaps is Plato's use of αἰών, seemingly for the first time in (philosophical) Greek apparently as an abstract noun or mass noun ("eon-ity") rather than as a count noun (a life, an eon, a [very long] time) (37d5, 38a7). But this transformation may have occurred simply so the term could serve here as the proper name for an Idea. Even within the discussion of time, then, the sense of the various terms for eternity will have to be determined by the project of which they are part.

I will first interpret Plato's view of time. I will suggest that when Plato says that the Demiurge makes time, he means that the Demiurge makes a clock, nothing more, nothing less.[4] Once this interpretation is established, we will be in a better position to interpret what he means by eternity as applied to various components of his metaphysics, including the Ideas, more especially because we will then be able accurately to formulate the question being asked when we ask after the temporal status of something. It will turn out that the Forms possess a *type* of timeless eternity, as opposed, that is, to possessing that cluster of notions, perpetual endurance, everlastingness, unchanging duration, and sempiternality.[5]

---

[2] *Phaedo* 78d4, 5, 6, 6–7, 79d2, 2; *Symposium* 211a1, 211b1, 4–5; *Republic* 479a2–3, e7–8; *Phaedrus* 250c3; *Statesman* 269d5–6; *Timaeus* 27d6, 28a6–7, 29b6, 52a1–2; *Philebus* 15b3, 3–4, 4, 59c3–4.

[3] Periphrastic uses of "eternal being" as a name for Forms: *Republic* 484b4, 485b2, 500c2–3, 527b5, 7, 585c1–2, 611e3; *Timaeus* 29a1, 3, 35a2, 37b3, 37e5, 48e6, 50c5; *Philebus* 59a7.

[4] To my knowledge Guthrie is the first person to see this point at all clearly. Guthrie 299–301.

[5] The following critics suppose that the Platonic Ideas are (merely) everlasting: Cornford (2) 102; Hintikka 6, 8; von Leyden 35–44, 52, and Whittaker *passim*. Cherniss (without argument), Vlastos (simply citing Cherniss), and Zeyl suppose that the Forms are timelessly eternal or in some sense "outside of time." Cherniss

Along the way I will be suggesting that Plato's views of time and eternity are more coherent than usually thought, and are not subject to a number of serious charges which have been leveled against them, though they are not entirely free of problems.

II. PLATONIC TIME

A. General Interpretive Framework

Time is a product of the Demiurge (37c8–d1, d5, 6; *Philebus* 30c). The Demiurge is usually interpreted simply as an arranger and orderer; he is viewed as introducing things of intrinsic worth, like beauty, into the phenomenal realm. Order, unity, and the like are viewed on such a reading as intrinsic goods. As we shall see, some critics try to assimilate Platonic time to this model by viewing it essentially as nothing more than a kind of order or arrangement, merely a more highly structured version of whatever existed before it as its materials (see section IIIB). This reading of the Demiurge is wrong or, at best, partial. Briefly, what the Demiurge does is to improve the world's intelligibility by introducing into the phenomenal realm standards, measures, or paradigm cases.[6] In this project he performs two sorts of actions. First, he makes in the phenomenal realm a perfect particular, a particular, that is, which corresponds precisely to the Idea of which it is an instance; thus the Demiurge accurately images in the instance the internal or proper attributes of its Idea (that is, what it is about the Idea that makes it the particular Idea it is). Second, the Demiurge invests the instance with the properties it will need to serve as a standard; thus he also images in the instance the properties which its Idea possesses *qua* standard or model.

In the Platonic metaphysics, then, we find two tiers of standards or measures: transcendental Ideal standards and standards immanent in the phenomenal realm (*Philebus* 57d). I suggest that in this arrangement the Platonic Ideas are to be viewed as standards *and as nothing else.* By this I mean that the properties which the Ideas have *qua* Ideas (that is, their formal, external or metaphysical attributes) can all be analyzed without remainder in terms of Forms being standards. There are two sorts of such formal properties. On the one hand, Forms have functional properties, that is, properties which the Forms have as fulfilling some role in Plato's scheme of things. These will be epistemological attributes (object

---

(1) 212, (8) 342–43, Vlastos (3) 408n3, Zeyl (2) xlii. Owen's interpretation seemingly has Plato wavering uncomfortably between these two views of the eternity of Forms (for discussion, see below sections IIIC and IVB).

[6] For details, see chapter 1, sections II, III.

of knowledge, that which makes definitions possible, etc.) and these functional attributes I suggest *result from* Forms being standards. On the other hand, Forms will have standard-establishing properties, properties, that is, which are *requisite to* their serving as standards. Uniqueness is one such attribute (*Republic* 596a6–b4, cf. 597c, *Timaeus* 31a4–7). I will argue that the sort of eternity which characterizes the Forms they also possess as a standard-establishing property.

Immanent standards, however, will not simply be standards in the way Forms are simply standards and nothing more. Immanent standards will be Janus-faced with respect to both their logic and status. For, on the one hand, with respect to their logic, they will be that by which other phenomena are identified and so these other phenomena may be viewed as instances of them, and in this way they will be like Forms in serving as means of identifying other things. Yet, on the other hand, immanent standards will themselves be instances of Forms; they will be perfect instances, corresponding precisely to their Form-standard, but they will be instances nonetheless.

Further, with respect to their status, though immanent standards may be invested with some of the standard-establishing properties of the Ideas (or substitute versions of these properties), they will nevertheless also possess all the characteristics that come along with being mired in the vagaries of the corporeal.

The Janus-faced nature of the immanent standards suggests that if we can analyze out of an immanent standard or standard case its component *as standard* from its component *as case* or *instance*, we will be able to learn something significant about how Plato conceives of its particular Form.[7] I shall now argue that time for Plato is one of these useful sensible images, which will help elucidate the nature of the Idea of eternity. (I take it that αἰών, *Timaeus* 37d5, 38a7, and even more obviously τὸ παράδειγμα τῆς διαιωνίας φύσεως, 38b8, and derivatively ἡ διαιωνία φύσις, 39e1-2, are all names of the Idea of eternity.)[8]

B. Time as a Sensible Standard

The view of time in the *Timaeus* is, I suggest, largely an elaboration of claims made first in the discussion of astronomy in *Republic* VII (528e–530d). It is here that the figure of the demiurgic God appears for the first time in the Platonic corpus. He is the craftsman of the heavens (530a6).

---

[7] See chapter 1, section III.

[8] This Idea will have whatever peculiarities are attendant on its being the Idea of a formal property of the Ideas. In this respect, it will be like the Idea of unity and those of the Great Kinds of the *Sophist* in which each and every Idea participates. 'Eternity', unity, and sameness will be said distributively of all Forms.

What he creates in producing the measured parts of the heavens—"the embroidery of the sky"—are called παραδείγματα (529d7). The context makes it clear that the term παραδείγματα here falls within its range of senses which cluster around 'exemplar'. It does not fall within the range of possible senses which mean 'example', 'illustrative instance', 'sample' or even 'parallel case'. For these immanent paradigms seem to be treated like blueprints in nature and function: "It is just as though we had come across plans most carefully drawn and worked out by the sculptor Daedalus, or some other craftsman or artist" (529d8–e3). Plato's analogy here refers to the custom of Greek sculptors to work from artificial rather than natural models. The sculptor in doing point and mallet work on, say, a marble horse did not look to a living horse as his model, rather he made an elaborate horse in clay (thinking of some living horse in the process) and then used the clay horse as his model for making a marble horse.[9] The artificial model will be Janus-faced in the way discussed above: it will be an image-instance of a model and yet will in turn be a model for other image-instances.

The examples which Plato gives of such demiurgically produced, earthly paradigms are temporal: night and day, months, years and "*other* stars" (530a7–8), all of which collectively are described as being material and sensible (σῶμά τε ἔχοντα καὶ ὁρώμενα, 530b3). By 'day', 'night', 'month', and 'year', therefore, Plato here is not referring to that which is temporally measurable (say, the duration of a festival), but rather is referring to the various planets and other celestial bodies treated as standards for measuring that which is measurable. In short, the demiurgically produced Janus-faced paradigms here are clocks. They are, however, not very good clocks. For they deviate from conditions of constancy which would enable them to be measured into each other (530a7–b3) and so enable them to operate in concert as measures. As they stand, they are incommensurable measures (cf. 531a1–3). There is no conversion factor between these measures. They do not stand to each other even as meter measures and yard measures stand to each other as measures of length. The reason for this incommensurability (we are to assume, and this is made clear in the *Timaeus*) is that each temporal measure (each clock) itself is changing even relative *to itself*.[10] Each fails to be constantly the same as itself with respect to that about it from which measurements are taken. If, analogously, The Standard Meter Stick were to change relative to itself, we would be unable to convert, say, fathoms and miles into meters and kilometers. That the temporal measures of the

---

[9] See Bluemel 36–39.

[10] *Timaeus* 38d, 39a–b; see Cornford *ad loc.* The circuits of the various planets are incommensurable (39c6–7) because they severally change rates of speed and even directions.

Republic are in this state of mutual incommensurability is directly attributed to their corporeal nature (530b3, participle obviously causal).

We have then in the *Republic* the two-tiered system of paradigms which I sketched above. (For Platonic Ideas as παραδείγματα, see *Republic* 500e3, 540a9, 592b2.) Within the *Timaeus*, the distinction between Ideal standards and immanent standards is drawn in the very opening section of Timaeus' discourse where he is laying out his principles (28c6–29a3).[11] Time in the *Timaeus* is one of these sensible, immanent standards, paradigms or measures.

The specific form which this immanent paradigm takes is that of a clock (or more accurately clocks) by which one can measure time (where 'time' is taken colloquially as that which is measurable about events and durations). A clock is a regularly repeating motion with some marker which makes possible the counting of the repetitions. Plato initially expresses this understanding of time viewed as a clock rather abstractly in his claim that time proceeds or (more precisely) revolves numerably (37d6–7, 38a7–8).[12] Stated more concretely, in the clocks which the Demiurge makes, it is the bodies of the planets—the wandering 'stars'—which are the clocks' markers and it is the planets' circuits which are the regularly repeating motions. Thus Plato says that the planets (viewed as moving bodies) come into being to define (or mark out, διορισμός) and guard numbers of time (38c6). They "define" numbers of time in that they are the clocks' markers. A perfectly homogeneous revolving disk, by contrast, could not serve as a clock, since even if its motion was completely regular, we would have no way of determining this or of counting its repetitions. We would in fact not even know (by sight at least) that it was moving. The circuits of the World-Soul, prior to the planets coming to be embedded in them (38c7–d1), are in fact like such a revolving disk in that the turnings of each circuit are not (at least to the senses) discernibly distinct. More specifically then, the planets define numbers in the sense of making these turnings discernibly distinct. The planets

---

[11] For discussion, see chapter 1, section IXA.

[12] Rémi Brague has noted that when at 37d6–7 Plato says, "This is what we call time," the antecedent of "this," which is masculine, is properly "number" (masculine), rather than "image" (feminine). Brague 66. In this case, the lines 37d6-7 should read, as against Cornford and others: the Demiurge "made, of eternity that abides in unity [i.e., the Idea of eternity], an everlasting image moving according to number. And this number is what we call time." Such a reading bolsters my contention that by time Plato is referring to the numerable circuits of the heavens viewed as measures. Brague's reading is supported, though it already has grammar on its side, by Plato's later identification of some of the moving heavenly bodies with time (39d1) and the fact that what we are said to learn from the study of time is number (47a6).

"guard" or "preserve" numbers of time in that their continuous wanderings each offer a single referent for numerable motions, and thus guarantee that the numerable motions are indeed repetitions, repetitions which are requisite for motions to serve as a clock. The regular pulses of a pulsar could be used as a clock, but the twinkling of stars at large could not: twelve pulses would mark a determinate amount of time; twelve twinkles could all be simultaneous. Repetitive motions require a single reference and the planets in their *continuous* wanderings invest Plato's celestial clocks with this feature.

When then Plato says that the planets collectively produce time (38e4–5) or more simply *are* time (39d1) he means by "time" something technical. He means that time is a clock, a clock by which we measure time, where 'time' here is used in a colloquial sense, as that about motion and rest which is measurable. The technical sense of time and this ordinary sense exist side by side through the text we are discussing (e.g., 38c2: ὁ δ' . . . χρόνον). Further, the two senses will seem to intersect and produce a second colloquial use when 'time' means that which is measurable *as actually having been measured,* that is, when a measurement has been taken *of* the measurable *from* the measure: "How much time did it take for them to get there? Three days' time." It is in this sense that 'time' appears in the description of the planets as "the instruments of time(s)" (42d5, 41e5). Here "time" would seem to refer to measurements taken. I do not see that Plato commits any fallacies of equivocation or other blunders which would suggest that he was confusedly running together the three senses of time: the measure or standard; the measurement taken; the measurable. The planets are measures from which the measurements of the measurable are taken. When then Plato identifies days, months and years with the various planetary circuits (39c1–5, also 47a4–6), he is treating them, as he did in the *Republic,* as that which measures or, more precisely, that from which measurements are taken. He is not treating days, months and years as that which is measurable or as that which has been measured. The days, months, and years are called "parts of time" (37e1–3) in the way we would call the various dials of a grandfather's clock parts of it. They are time's types or species.

In sum then, the time which the Demiurge creates is an immanent sensible standard which has the specific form of being a clock (or composite clock), a conspicuous measure of change.[13]

[13] The following is at least as likely the right rendering of 39b2–4 as Cornford's:

> In order that [the heavenly bodies, 38e5] in their various relative speeds should be [i.e., should serve as] a sort of conspicuous measure and [in order that] the things regarding the eight planetary circuits should be con-

There are a number of things which demiurgically created time is not. First, the Demiurge does not invent temporal succession itself; he does not produce events *simpliciter* or produce them as having transitive, asymmetrical, irreflexive relations. This, though, has been denied. It has been suggested that before the Demiurge made the universe and time, the content of Platonic space was a sort of muddy blur without discernibly distinct bodies or events.[14] The temporal status of this alleged condition has been called "some type of durational spread."[15] This view though does not jibe well with Plato's description of the pre-cosmic chaos (52d–53c, cf. 30a). The emphasis of the description is on diversity of appearance and plurality of motions (especially 52e1–2; for bodies in the pre-cosmos, see 50b6).[16]

Second, and more importantly, the time which the Demiurge creates is not accurately characterized as "uniform and measurable time-flow."[17] If time in Plato's technical sense is fairly described as a clock, then time is that by which or from which the measurable is measured and is not that which is measurable. *Qua* measurable, there is no difference between the events of the pre-cosmos and the events of the ordered world. The Demiurge makes no alterations in events themselves in order to make them capable of being measured. He simply introduces a standard so that that which is measurable may actually be measured against it. The changes

---

> veyed [i.e., the information they hold should be made apparent to us], the Demiurge kindled a light in the planet we call the sun.

The sun just by itself is not the measure which the Demiurge creates nor is it that from which alone, as an object of cognition, we learn numbers (39b6); rather all the easily conspicuous planetary circuits are sources for learning numbers (47a). Therefore, it seems likely that εἴη (39b2) is a predicative rather than an existential usage; its subject, like the antecedent of ἄλληλα (b3), is the many preceding neuter plurals which refer to the heavenly bodies (collectively or in groups) going back to σώματα (38e5). Παρά (39b7) means "from" but not necessarily "about," as in "we learned from you that . . . . ." The sun, through its light, is a source of knowing about all the heavenly bodies. AY texts retained at 39b3: καὶ τά.

[14] Robinson (1) 97&nn. I take it that Zeyl has something similar in mind when he writes: "By setting the heaven in motion the Craftsman creates time, a supervenient aspect of that motion." Zeyl xlii n80.

[15] Whittaker 137. Vlastos correctly points out that if there were no temporal 'passage' prior to creation, the Demiurge paradoxically enough in creating time would make the world "far *less* like the Ideas than it otherwise would have been." Vlastos (3) 411, Vlastos' emphasis.

[16] For a detailed analysis of the pre-cosmos, see chapter 6 below.

[17] This is Vlastos' formulation of the Demiurge's accomplishment. Vlastos (3) 410.

wrought by creation are all on the side of the standard and not on the side of what gets measured: orderly (or circular) motions and disorderly (or rectilinear) motions are equally measurable. Further, the regularity or uniformity which the Demiurge introduces into the world (at least as regards time) is not meant as a transformation, working over, or developed state of the transitive, irreflexive, asymmetrical temporal relations which exist prior to the world's formation; rather regularity in the form of regular repetition is a requisite for a clock to serve as a measure of those temporal relations (which may hold of either orderly or disorderly motions). Plato makes the point that time is not a developed state of the measurable in a literary way by having the physical motions of the planets or "instruments of time" derive entirely from the psychic motions of the World-Soul, in which the planets are embedded, rather than having their motions be alterations of physical motions taken over from the pre-cosmos (38c–e).[18]

Critics' confusion of time the measurable with time the generated measure has precipitated the charge that Plato's account of time is fundamentally incoherent when set against the general metaphysical backdrop of the *Timaeus*.[19] In particular it is claimed that the presence in the pre-cosmos of temporal relations which possess transitive, asymmetrical, irreflexive *order* is incompatible with the view that all *order* is the product of the Demiurge. However, this claim about the Demiurge is false or at least overly broad. What is true of the Demiurge is that all order and arrangements *which* manifest purpose and proximately serve functions are purposeful and functional as the result of his doings. Transitive, asymmetrical, irreflexive relations *may or may not* manifest purpose and serve functions. Take the relation to-the-left-of as an example of a relation exhibiting such transitive, asymmetrical, irreflexive order. If I arrange books on a shelf in some way that is useful to me, say, alphabetically or by date of publication or by size or by call-number, then the various to-the-left-of relations which hold between the books will manifest purpose and serve a function. If, however, I shelve books just to get them conveniently out of my way and pay no attention to which book is where on the shelf, then the various to-the-left-of relations among the books will not manifest purpose or serve a function. In the pre-cosmos such relations and indeed all regular or law-like features operating in (and even

[18] For a different reading of the relation of planetary motions to the circuits of the World-Soul, see Crombie. Crombie 208. Crombie's suggestion that the Demiurge creates an 'absolute time' which is independent of planetary motions seems to be defeated by 38e4–5 and 39d1.

[19] The following charge of incoherency is laid by Vlastos (2) 389–90, (3) 413–14.

producing) the flux of phenomena will lack purpose and function.[20] In the demiurgically ordered world some will and some will not be purposive, depending upon whether they are taken over and incorporated into one of the Demiurge's projects (46c–47d). There is then nothing the least bit incoherent with Plato's commitment to pre-cosmic temporal order taken in conjunction with his view of the Demiurge's nature.

C. *Telling Time*

On my interpretation of the Demiurge's actions, he introduces standards into the phenomenal realm in order to improve our cognitive capacities. What specific form does this improvement take?

For Plato, the presence of standards in the phenomenal realm has the consequence that if we have just the phenomenal realm present to us and not also the Ideal realm, nevertheless we are still able to make the sorts of judgments of measurement which can *only* be made by measuring against a standard. Judgments of measure made by reference to standards, though, are not the only sort of metric judgments which we make. We also make judgments based on merely relative measurement. By "merely relative measurement" I mean that ability which Plato claims we have, usually by direct sensory inspection and in any case without reference to measures or standards, to say that, on some scale of degrees, one thing is greater than another, or even possesses the same degree as another, but which does not entail the further ability to say what degree on the relevant scale of degrees either thing possesses. We would have the latter ability only if we could appeal to a standard or measure (*Statesman* 283c–285a). Plato gives a number of examples of scales of degrees tokens of which are subject both to merely relative comparison or measure and to measure against a standard: length, breadth, thickness or density (B text), swiftness (T text) and number (i.e., integers) (284e). We may say that one person is taller than another by direct visual inspection without appeal to a standard, but will not be able to say *how much* taller the one is than the other. With the aid of a meter stick, however, we are able to *identify* precisely how tall each person is and then, derivatively, to say precisely how much taller the one is than the other. Actually these latter abilities do not attach to the use of all standards. They would not hold true of the use, say, of a standard sepia patch. With it we would be able to say that one individual color token is more closely sepia colored than another, but would not be able to identify precisely the color of the tokens or determine precisely their difference. But for those scales of de-

[20] For law-like features in the pre-cosmos, see especially 52e–53a, 57a–c, 57e–58c, and for an analysis of these passages, see chapter 6 below.

grees for which that which is measured is numerable, the precise identification and comparison of tokens of the scale's degrees will be made possible by the use of standards. All of Plato's examples of measures fit this latter model (cf. *Philebus* 24c-d) and temporal measure also belongs to this class of measures (*Philebus* 30c6).[21]

In the case of measuring against a standard, the individuals measured are not measured directly against each other as they are in the case of merely relative measure. Rather they are first each measured severally by being identified each by reference to the standard for their scale, and then the measures which have been taken severally are compared through our knowledge of the properties of the natural number line.[22]

It appears that Plato wishes to exclude from his schema of judgments the sort of use of standards which measures precisely the difference between two tokens on some scale by making a single direct measurement of the 'gap' between the two tokens, for example, by laying two boards side by side with their termini juxtaposed at one end and then using a measuring rod to make a single measurement of the stretch by which the one board exceeds the other. The discussion of the Forms Tallness and Shortness in the closing argument of the *Phaedo* makes it clear that judgments of comparison which appeal to Forms are viewed as entailing first a separate identification for each relata in the comparison. The judgment of comparison is then derived from the separate identifications (100e–101c, 102b–e). Thus, Simmias is taller than Socrates (on an account that appeals to Forms) because of 1) the way tallness in Simmias and shortness in Socrates are respectively identified by reference to Tallness and Shortness and 2) the way Tallness and Shortness stand related to each other. It is not by appealing to some precisely and directly identifiable feature of the difference between the items compared (say, the difference of a head) that the judgment of their difference is to be made (100e–101b). For Plato, identification is prior to comparison, when one uses standards to make comparisons.

If we turn to temporal measure, we see that in creating 'time' or a clock not only is the Demiurge not creating temporal succession but also he is not even creating the means by which we may perceive and make judgments of sheer succession. Judgments of raw temporal succession will be relevantly similar to raw judgments of greater and less. The Demiurge provides the means of making the sorts of temporal judgments which one can only make with the use of a standard. And even

---

[21] In the *Philebus*, the imposition of measure on the more and the less is equated with the establishment of determinate number, definite quantity (ποσόν) (24c3, d3, 5, 6, 7).

[22] For Plato on the natural number line, see chapter 15.

without appeals to a clock, we can make judgments of sheer succession, that is, judgments that one event occurred before or after another, earlier or later than another. Thus, if I snapped the fingers of one hand, paused, and then snapped the fingers of the other, anyone in the room who was paying attention could say which of the two events occurred first without resorting to making references to a clock. But the two events could also be measured by appealing to a clock taken as a standard for measuring events and durations. Each snap could be severally identified by being assigned a date through a reference to the clock, say, 6:31 and 6:32, and given a temporal ordering as the result of 1) this matching of each event severally to one of the clock's markings and 2) a subsequent comparison of the measures taken. With the use of the standard of temporal measure, we will know not only that the one snap occurred earlier than the other, but also we will know when each occurred and how much later the one is than the other (in our example, one minute). The crucial component of identifying tokens, when measuring with the aid of temporal standards, is the assigning of determinate dates to events, that is, specifying for an event a date type from the scale of possible date types. Any particular date type (say, 6:31 pm) may have any number of events as tokens (e.g., births, train departures).

Our ability to make comparisons (both relative and precise) of *lengths* of duration derives from our ability to make (relative and standard) judgments of earlier and later. Relative judgments of earlier and later require the operations of memory in a way which relative judgments of, say, lengths of directly and jointly perceived objects do not. When our capacity for memory is combined with even rudimentary imaginative skills, we are able to suppose or imagine that two remembered events had started at the same time; the longer of the two will then be the one that ends later than the other in our imagined scenario. This capacity to judge relatively lengths of duration is analogous to our ability to compare relatively the size of two mounds which are not both ever immediately present to us. To make precise comparisons of two lengths of duration with the aid of a temporal standard entails making four designations of dates, specifically the starting date and the ending date of each length compared; a derivative determination of the precise length of each is then made, and then the precise difference of durations will result simply from comparing the number of each length. If we keep in mind that for Plato the identification of terms precedes the comparison of terms when one uses standards to make comparisons, then it is clear that Plato does not suppose that the celestial clock is to be used in the way we use a stopwatch, that is, by making a precise measure of comparison with a single direct measure of the stretch by which one event extends beyond another. At best we would be speaking elliptically if we said that in finishing the race Simmias is earlier than Socrates by a minute.

D. Past and Future and their Relation to the Formation of the Cosmos

When Plato calls the 'it-was' and the 'it-will-be' εἴδη (forms, kinds, types) of time (37e4), I suggest that he is referring to time in its sense as that which has been measured, that is, as measurements taken. The past and future, as specifications of time measured, are thus distinguished from the earlier mentioned 'parts' (μέρη) of time (37e3), the various formally and numerically distinct standards which make up the composite celestial clock.

Many critics have thought Plato's claim that the Demiurge brought past and future into being stands in such blatant contradiction to the suggestion of Timaeus' narrative that there was a time prior to the formation of the heavens that the discussion of time must in part be signaling that the narrative is not to be read literally—that there indeed was no initial act of world formation on the part of the Demiurge.[23] I suggest there is no such lurking contradiction which will generate a motivation for a non-literal reading and that indeed Plato will make more sense if he is not committed to such a motivation.

The only distinction between judgments of merely relative temporal comparisons of earlier and later, before and after, and judgments of past and future is that the latter are indexical or token-reflexive expressions, that is, any statement about the past, present or future directly makes reference to the very act of making the statement. Judgments of past, present and future are like other temporal comparisons only with the proviso that one of the relata of the comparison is the very act of judgment making on the part of an actual or hypothesized speaker or thinker. Thus "*x* is past" goes over into "*x* is earlier than (the making of) this statement," "*x* is present" goes over into "*x* is simultaneous with this statement" and "*x* is in the future" goes over into "*x* is later than this statement." Such comparisons can be made as the result of merely relative comparison, just as are any judgments of earlier and later which do not require appeals to standards. However, judgments of past, present, and future may also be made by reference to a clock, in which case we may make in addition to the raw judgment of past, present, and future, a judgment of just how much in the past or future something is and we may identify when the present is. Thus if, as I speak, it is 6:30 pm and if a train departs at 6:32 pm, the departure is in the future and precisely two minutes in the future. The past and future which the Demiurge's creations allow us to discern are of this latter sort. But the before and after,

[23] For a review of the literature and an attack on non-literal readings, see Vlastos (3). For a defense of a non-literal reading, against Vlastos, see Tarán. For a defense of a literal reading, see Zeyl (2) xx–xxv. Johansen verges upon a non-literal reading, but wavers. Johansen 89–91.

past and future which belong to the narrated events of the *Timaeus* are of the sort determined by merely relative comparisons, which are made without the aid of a temporal standard. As a result, there will be no contradiction between Plato's claiming, on the one hand, that the Demiurge made past and future and, on the other hand, that there was a time past prior to his creations.

It has been denied though that it makes sense to speak of different sorts of before and after, past and future. It has been claimed that past and future are species of time and stand to time as red and blue stand to color, or as even and odd stand to number, such that it would be as silly to say that there was a past event when there was no time as it would be to say that there were blue and red things but no colored things and even and odd things but no numbered things.[24] These analogies however are misleading. Past and present do not stand to the time which the Demiurge makes, i.e., to the celestial clock, as blue and red stand to color. For saying that the existence of events as past and future depends upon the invention of a clock is like saying that determinations of longer and shorter must await the invention of The Standard Meter Stick, or that if longer and shorter are measured by The Standard Meter Stick, they cannot be measured in any other way, say, by the use of a yard stick. Past and future are merely types of the actually measured. Their existence, therefore, is not contingent on the existence of any particular mode of measurement.

Therefore, those critics who would read the discussion of the production of time in such a way as to perceive a contradiction blatant enough to motivate a non-literal reading of the production of time and the heavens can succeed only at the high price of accusing Plato of being conceptually confused. They must accuse Plato of failing to distinguish a standard or measure of time from the measurements which might be taken from the standard.

III. PLATONIC ETERNITIES

In consequence of the preceding sketch of the nature of Platonic time and temporal judgments, we are now in a position to analyze the temporal status of various parts of the Platonic ontology.

A. The Phenomenal Realm in General

First, in making 'time' the Demiurge does not in fact change the temporal status of the phenomenal realm. Whether one reads the creation

---

[24] This is Tarán's strategy for establishing a non-literal reading of the *Timaeus* based on its discussion of time. Tarán n59.

story of the *Timaeus* literally or not, the phenomenal realm is sempiternal independently of anything which the Demiurge does to it. By "it is sempiternal" I mean that from any event in it we can in every case indicate an earlier and a later event (established even just by relative, i.e. non-clock-dated, comparison), or, more informally, mean that it exists through all time *by* existing at all times (possible dates) infinitely into the past and infinitely into the future. And this sempiternal 'condition' of the world obtains unchanged even after the Demiurge's interventions into the world. What the Demiurge does is to introduce stability or uniformity into the world with respect to some motions, so that some of the world's motions have sufficient regularity or even-repetitiveness such that they in turn may serve as standards for measuring its other motions.

In making time or more precisely in making the means to tell time with clocks, the Demiurge does not alter the temporal-ontological nature and status of the world, rather he establishes conditions so that things which have that nature and status become more intelligible. The Demiurge's mission here, as elsewhere, is epistemological in purpose (so explicitly 39b5–c1, 47a1–7).

B. Platonic Time: The Celestial Clock

Second, the celestial clock in consequence of its Janus-like nature as a sensible standard will have a two-fold temporal status. When it is considered under its description as just any phenomenal object, its temporal status is that of a contingent and partial sempiternality. *As determined by merely relative measure,* it will exist at all possible dates from its formation on. If one picks any event in the formed universe, some motion of the celestial clock will be prior to it, simultaneous to it and after it. The celestial clock possesses a *partial* sempiternality because it does not exist at all possible dates into the past. And its sempiternality is *contingent.* For Plato entertains as a physical possibility the dissolution of the celestial clock back into a state of chaos at some future date (38b7, cf. *Statesman* 273c–e).

Under its description as a phenomenal object, the celestial clock is subject to merely relative measure just as is any other phenomenal object. But when it is subjected to merely relative measurement, it is not being treated as a standard for measurement. An illustrative analogy: if I treat The Standard Meter Stick in Paris not as a standard from which measurements are taken but rather as any other extended object, I may say that it is shorter than the length of my desk. But this is to subject it to merely relative measure, and is not to treat it as a standard for measure. So too, I may say that Plato's celestial clock did not exist prior to the ordered heavens (38b6–7), but did exist prior to the sinking of Atlantis. To make these sorts of claims, though, is to fail to treat the celestial clock as a standard from which measurements are taken; it is to treat it like any other thing in the phenomenal realm. The celestial clock *qua* phenomenal object may be said to last forever; and this indeed is how Plato describes it: διὰ τέλους τὸν ἅπαντα χρόνον (38c2).

If, however, we consider the celestial clock under its description as a standard, it will have quite a different temporal status. *Qua* standard, the celestial clock, the Demiurge's created time, will have a temporal status which might be described colloquially as a timeless, non-durational eternity or a subsistence outside of time. What I mean precisely by this is that it does not fall within the category of things to which it is possible to assign dates and of which it is possible to measure durations. This eternity is not to be confused with sempiternality; indeed it is incompatible with sempiternality as defined above. For something to be sempiternal is for it to exist at all possible dates. For something to exist timelessly is not conversely for it to fail to exist at all dates (like say, unicorns), but to be such that the questions "when is it?" and "how long did it last?" are inappropriate to ask of it. This sense of eternity has nothing to do with lasting, even lasting quite a while. Many different types of things may have this status and for many different reasons.[25] I suggest that Platonic standards will have this status as part of their nature as standards.

Quite generally, standards *as what we use to measure change* are incapable of being thought of *either* as changing *or* as not happening to change. We know that something other than a standard is either changing *or not,* in either case, by reference to a standard. Thus we determinately tell both that a child has been growing and that an adult has not—both—by reference to a standard for length. But this means that the standard itself is neither changing nor just happening (like the adult) not to change. It is in this sense as neither changing nor happening not to change that standards "stand outside of time."

---

[25] Numbers on most intensional accounts will fall outside of the temporally measurable. Aristotle's arguments in *Metaphysics* VII, 8 look like they are an attempt to place both form and matter outside the category of the temporally measurable, rather than to view them merely as sempiternal.

One sort of timeless eternity which the Platonic Ideas do not possess is an eternity *totum simul,* an eternity, that is, in which an eternal object, like Boethius' God, contains within any 'moment' of its existence the entire history of the world, on an analogy with the way in which a tape recording sitting on a shelf possesses within itself, all at once, a full-length concert. Nor, conversely, is Platonic time, as is claimed on some Neoplatonic accounts, a projection or emanation from such an eternal object, as though time were, in our analogy, the concert as produced when the tape recording is played. Kneale finds hints of eternity *totum simul* in the *Timaeus*. Kneale 93–94; against his view see, Whittaker; against Neoplatonic views, see Cherniss (5) 234–37.

It is true that we can suppose that virtually everything is capable of change *when* we make this assessment by means of merely relative measurement. And sometimes we assign change to standards in this way: "My watch stopped," "My measuring tape shrank in the wash," "Relative to the tree's growth my yardstick became smaller." But in these sorts of cases we are not treating the thing measured in this way, whatever thing it is, as a standard, rather we are treating it merely as any material object. A sign of this is that the only occasion on which it is natural in ordinary discourse to predicate (change) of standards with respect to their status as standards is in fact when the standards have broken and so have reverted to being just any material objects. When they are working, we do not make predications of them; we use them to make predications. Standards *qua* standards are by the definition of what it is to be a standard not in the category of things which either change or abide, even endlessly abide. We cannot say they abide or stay the same as themselves; for we have no way to determine this other than by merely relative measure, and to treat them as mere relata in relative measures is to cease to treat them as standards.

We can see that this is true in the specific case of temporal measures like Plato's celestial clock. For it makes no sense to ask of a clock whether it exists at 6 o'clock; for "to exist at 6 o'clock" means to exist when the clock's pointer is aimed at the numeral six, and "6 o'clock" here is a proper name (assuming the clock is accompanied with a calendar).[26] The pointer being aimed at 6, we might say, is the essence of 6 o'clock-ness. Other things are said to exist or occur at 6 o'clock or be 6 o'clock events by reference to it. But it itself is not a 6 o'clock event. It is the meaning of 6 o'clock, what it is to be 6 o'clock. Further, it makes no sense to ask how long it takes for the pointer to circumscribe the clock's dial. We may col-

---

[26] The calendar aspect of the celestial clock is provided simply enough by the fact that the series of integers is infinite (see *Parmenides* 144a). Cherniss has suggested that the reference to "the perfect/complete number of time" (39d3–4) shows that Plato cannot be thinking of the numbers of time (referred to elsewhere in the discussion of time, notably at 37d6, 38a7) as the open-ended series of integers. Cherniss (5) 234n34. However, the complete number here is an aspect of the 'complete year' (39d4), and though both these notions are somewhat obscure, I suggest that the complete year is a unit of time which is a parasitic compounding of other units of time. "The complete year" is relevantly similar to the expression "a month of Sundays." "The complete number" then will be relevantly similar to "about thirty" in the sentence "a month of Sundays is about thirty weeks." The compounded Platonic 'year' and its enumerated sub-units may of course like our 'month' of Sundays, be repeatedly marked off without end, each subsequent repetition being assigned a different, though sequential, integer which will provide its distinguishing (proper) name.

loquially say that it takes the pointer an hour, making this assessment by the merely relative comparison of it to other things which we, through repeated acquaintance, know typically to take about an hour, but we may not say it takes an hour in a strict way, that is, by reference to a measure. For what it means for something (say, a television program) to be an hour long is just that its starting and stopping dates are marked and measured by one of the pointer's circuits.

Therefore, if we say that it takes one hour for the pointer to circle the dial, we do so by definition or as an identity statement. The Demiurge, in making part of the world into a giant clock, removes that part of the world *in its aspect as a standard* from being the subject of judgments of dates and duration.

It is in this sense then as failing to fall within the category of things which are subject to temporal judgments that the time which the Demiurge creates is said to be eternal (37d7). And in this respect it differs substantially from the ordered heavens which are merely long lasting. Some critics have tended to confuse Platonic time with the sheer orderliness of the world, and so have tended to construe the eternity of Platonic time as a sort of permanence. Thus, for instance, we hear: "In short, time is the rational aspect of orderliness in the phenomenal realm by which the flux of becoming can *simulate* the eternity of real being."[27] And hear: "[Time for Plato is] the all-inclusive system of orderliness in nature."[28] However, Plato would not repeatedly say that the ordered heavens and time came into being simultaneously (37d5–6, 38b6), if he in fact meant that they were one and the same thing, thus rendering otiose any claim about their simultaneity. Further, there is no suggestion that all ordered objects of the Demiurge's formation (e.g., souls, human bodies, plants, metals) are in some sense components (either formal or material) of Platonic time viewed as a system. And what order Platonic time does possess does not even constitute its final characterization or essence. Rather, its orderliness, which has as its forms even-repetitiveness and numerability, is simply a necessary precondition for it serving as the standard it is. Platonic time then is not an eternal object in some sense as containing, encompassing, or possessing as parts that which is fleeting, a system which, while its parts come and go, itself abides. Nor is it in any other sense eternal simply in consequence of being an ordered entity.

C. *The Ideas*

Plato calls the model for Platonic time αἰώνιος (37d3) and he also calls Platonic time itself αἰώνιος (d7); the eternity of Platonic time, however, is not eternal παντελῶς (completely, unqualifiedly, simpliciter, d4), for it is

[27] Cherniss (5) 236–37.

[29] von Leyden 52, cf. 43.

a generated object (d4). The proviso that Platonic time is not eternal παν-τελῶς has been taken, on one account, as implying that the two different occurrences of the term αἰώνιος here must have strongly equivocal senses and, therefore, as entailing the inference that if αἰώνιος as used of the Ideas means "timelessly eternal," then the term as applied to time must have a different sense, with sempiternality or everlasting duration being the most likely alternative.[29] But this is an odd critical stratagem. For if I say "*x* is F, *but* not completely F," the statement's contrast would be fatuous if the two occurrences of "F" were equivocal. I will be making a poor joke if I substitute "red" for the occurrences of "F" and in the first case mean "of a far left political persuasion" and in the second case mean "sunburned".

Alternatively, the proviso has been construed (in a way which avoids this difficulty) as a reference to the canonical formulations of the theory of Ideas in the middle dialogues, which, on one reading of the theory, view a Form as having unqualifiedly the property of which the Form enables the identification in other things, whereas the other things, which participate in it, both possess and do not possess the property in different respects. Thus in the case at hand, the relevant Idea is unqualifiedly stable or permanent, but time, its instance, is "both stable and unstable: the pattern is stable, but it is a pattern of change."[30] If I am right so far about the eternity of Platonic time, in that order and stability are not what constitute Platonic time as eternal, we can rule out this reading of the proviso.

I suggest, as both Plato's word choice and his doctrine of measures would indicate, that the Forms and Platonic time are eternal *in the same sense* and that this sense is that of falling outside the category of things of which it is intelligible to make temporal judgments of dates and durations. The Forms will possess this sort of timeless eternity in an unqualified way as the result of their *status* as being standards and nothing else, whereas the celestial clock will have this property only *qua* standard. It will not, however, have this property unqualifiedly. For it will also be subject to temporal judgments in consequence of being a generated object, part of the phenomenal realm. The proviso then refers not to the internal or proper attributes of time (i.e., those which distinguish it from other phenomenal objects)[31] but to the external or metaphysical attributes of Platonic time: completeness designates ontological status.[32] Further, the incompleteness which prevents time from being unqualifiedly

[29] Cherniss (1) n350, cf (1) 212.

[30] Owen (2) 334.

[31] This is Owen's view.

[32] Insofar as the proviso makes a point about status and insofar as it is in fact a reference to the canonical formulations of the theory of Ideas in the middle dialogues, the various descriptions of the Ideas in the middle dialogues as "what

eternal takes the form not of a lack or absence, but of an accretion, as in the way superfluous stone on a partially finished marble sculpture marks it as incomplete or as a husk marks a seedling as incomplete. The accretions in the present case are the metaphysical ones which attach to an object's presence in the realm of becoming.

Now admittedly at first inspection, Plato's elaboration of the nature of the eternity of the model (37e4–38a8) might not suggest such an analysis. The argument for the eternity of the Forms first appears to be: the Forms are unmoving (ἔχον ἀκινήτως. 38a3); but, the past and future are motions (κινήσεις γάρ ἐστον, 38a2), and for this reason they are inapplicable to unmoving Forms. If this is Plato's argument, then he is profoundly confused. For the Forms, on this account, will (merely) be permanent or stable, but stable, curiously enough, without being in time; and yet "being stable" means being the same from time to time, and so the Forms are not after all stable in any legitimate sense.[33] If Plato really believes that past, future, and more generally all relations which are constitutive of temporal order count as changes, if, that is, Plato seriously believes, as we might say colloquially, that time flows,[34] then Forms will at best merely be permanent or everlasting, but permanent in some overly contrived sense. To say, in consequence of the argument construed as above, that a Form "only is," but "is" without past and future is merely an *ad hoc* expedient which clarifies nothing.[35]

However, in initially calling past and future "motions" (38a2), Plato is speaking casually or elliptically. He expresses himself more fully and accurately later in the same sentence, where it turns out that past, present and future (38a4–5) are not viewed as motions but as properties which supervene on motions: they are things "which becoming attaches to sensible movings" (38a6). The past, future, and even the present, on this fuller account, are those temporal properties which we describe in judgments of measure, either relative measure or measure against a standard. The passage is claiming that there is some sense in which these judgments, including judgments of being-in-the-present, are completely inapplicable to the Forms. The "only is" which is applicable to the Forms is

---

completely or purely is" (*Republic* 477a, 478d, 597a) are using (complete) existential senses of the verb "to be" and not predicative or veridical senses (so also *Phaedo* 66a, 67a), and so further the formulation of the theory of Ideas mentioned in the previous paragraph is wrong.

[33] This is Owen's understanding of Plato's argument and its (alleged) confusion. Owen (2) 335.

[34] Plato could easily have spoken here of time as moving or flowing, if this is what he intended; note πορευομένου τοῦ χρόνου ("time advances"), *Parmenides* 152a3–4.

[35] So Owen (2) 335.

not a "timeless" present illegitimately rent from its logical connections with past and future—the only connections which will make 'the present' an intelligible notion. The Forms will have the requisite sense of αἰώνιος, not being subject to temporal judgments, including judgments of the present tense, *if*, as I suggest, they are viewed as standards and nothing else.

Plato admits that in casual discourse we indeed are, from our perspective in the phenomenal realm, prone to apply "past" and "future" to the Forms, but he claims we would be wrong to do so; we do so only unthinkingly (37e4–5). Plato, I suggest, is here acknowledging our human tendency to take *anything* and everything as subject to measurement, either merely relative comparison or even measurement against a standard. We may casually ask, "Did the Ideas exist prior to the making of this statement?" that is, "Are the Ideas in the past?" or even ask "Did the Forms exist at 6:30 pm?" But to do so, in the first example, is to treat the Forms as a mere relatum in a merely relative comparison, rather than to treat them as that from which we take measures and, in the second, is to treat them as instances of standards rather than as standards themselves.[36] To treat Forms in these ways is to treat them as completely superfluous metaphysical baggage, since on the one hand, we are capable of making merely relative judgments independently of them and on the other, they cease even to be standards, but rather are instances of standards. If Forms are standards and nothing else, then these sorts of questions, calling for temporal judgments as answers, cannot legitimately be asked of them. The sort of temporal comparison—between a datable event and a timelessly eternal object—which Plato wishes to rule out of court as a conceptual confusion is exemplified in Christ's remark "Before Abraham was, I *am*" (John 8.58).

Forms are that by which we make precise judgments of comparison and difference in other things, by first assessing and identifying the other things severally against a Form as standard. In their role as standards, the Forms are not in the class of items of which they make possible the comparisons and differentiations. As not subject (*qua* standards) to judgments of similarity and difference, they are not in time at all. Plato is explicit about this in the *Parmenides*, where as a direct result of anything's failure to be subject to judgments of same and different, likeness and dissimilarity, it is correctly claimed that the thing (whatever it is) cannot be in time in any way:

---

[36] To try to measure the Idea of eternity by using the celestial clock as a standard is like trying to measure The Standard Meter Stick in Paris using my balsa wood measuring stick as a standard of measure. In this circumstance, The Standard Meter remains a standard in name only, as a quirk linguistic history, rather than because of its nature as a standard.

> If it is the same age with itself or another, it will have sameness of time and likeness; and we have said that the one has neither likeness nor sameness. We also said that it has no unlikeness or inequality. *Such a thing* (οἷον) cannot, then, be either older *or younger* than, or the same age with, *anything* (τι). Therefore the one cannot *be younger* or older *than*, or the same age with, *itself or another.* We may infer that the one, if it is such as we have described, cannot even be in time at all (140e3–141a6).

In our *Timaeus* passage, the Forms are described as never becoming "older *or younger* through time" (38a3–4). This curious formulation has been universally and correctly viewed as an allusion (either as foreshadow or echo) to the *Parmenides* passage just quoted. The reasoning of the *Parmenides* passage is sound and it applies to each Form, since each Form in being the particular standard it is, fails to be a candidate for a role as a relatum in comparisons. The Forms then will be like the celestial clock *in that* they as standards do not exist in time at all; they do not exist at any dates and are not within the category of things which are subject to temporal judgments. To construe the Forms either as changing or as merely permanent is to cease to view them as standards and to abolish the special cognitive capacity of assessing things against standards in favor of only being able to make merely relative judgments.

When then Plato describes the Forms as ἔχον ἀκινήτως, he means not that they simply happen not ever to change, but that they indeed are not capable of change as falling outside of the class of things which change or fail to change.

Stated more precisely, Forms fall outside the class of things which undergo *alteration* or do not undergo alteration. By "alteration" I mean that sort of change which can, as a logical possibility, be thought of as taking place in a thing independently of its relations to other things. For Forms stand to the phenomenal realm in such a way that they will have *relations* which come and go. Thus a Form may be known at one time but not another, or may be imaged or instantiated at some place at one time but not another. The Forms, then, will be subject to merely relational changes, changes, that is, which do not entail that Forms alter in any way. A merely relational change though is parasitic upon an actual alteration in the object in relation to which a thing undergoes a relational change.[37] The merely relational changes of a Form will be parasitic upon alterations in the phenomenal realm. The knower or the locus of participation undergoes not a merely relational change but an alteration. Further it is alterations or the lack of them that are the proper subjects of judgments

---

[37] For an analysis of "merely relational changes" or what are sometimes now called "Cambridge changes," see Lombard 63–79; for their entirely parasitic nature, which has led philosophers to consider them as non-real or bogus changes, see especially 74–76.

of dates and durations. Such judgments apply derivatively to the merely relational changes that are parasitic upon alterations. But if in consequence of this parasitism we try to extend these judgments to the Forms, which neither alter nor do not alter, we do so again only by illegitimate merely relative comparisons. If, for example, we try to derive the claim that a Form existed for six minutes from the claim that it was known for six minutes, we do so by merely relative comparison.[38]

Do Forms, then, have histories? The answer is yes and no. In a very weak sense of "history," a thing has a history if something is true of it at one time but not at another. On such a view of what counts as a history, the drying of ink on this page constitutes a history of the Eiffel Tower. For at one time it is true and at another false that an object with the very distance and direction from the Eiffel Tower possessed by this page is wet with ink. The alteration in the ink produces a merely relational change in the tower, which in turn is sufficient for it to have a "weak" history. Similarly, as the result of undergoing such merely relational changes as being known and being imaged, Forms will have histories in this utterly trivial sense, and there is nothing incompatible between a thing being timelessly eternal and having a history in this sense. How-

---

[38] Patterson's account of the Forms' eternity as 'an absolute immunity to change' simply fails to take into account that Forms will *on anyone's account* undergo merely relational changes, since they come to be known and participated in. Indeed the very flux of the phenomena guarantees that the Forms undergo these changes. It is hard to know therefore how Forms for Patterson (without some qualification or under some description) can be 'utterly immune to becoming'. Patterson (3) 36, 37.

Patterson wishes to distinguish the temporal status of Forms from that of close contenders by claiming that Forms in principle cannot change whereas other things even if they happen never to change nevertheless acquire histories simply from being capable of change—like an "object frozen by God forever." Patterson (3) 36. If, however, it is merely *alteration* from which Forms for Patterson are utterly immune, their immunity will not prevent them from acquiring histories *in exactly the same way* as those objects which might alter, but indeed do not alter, acquire histories—some of which on Patterson's account are even to be found in our world. The difference between that which can but does not alter and that which does not alter because it cannot is simply irrelevant to the acquisition of temporal properties. For temporal properties themselves are merely relational changes that result from alterations in the standards of time: alterations that make moments discernibly distinct, say, 6 o'clock and now. Whether something in principle does not alter or simply happens not to alter is irrelevant to determining whether it exists at 6 o'clock yesterday.

The way out of this difficulty is to recognize that Forms are essentially standards and, *qua* standards, are not even subject to merely relational judgments and changes.

ever, if we take history in the more usual sense as the course of actions and passions, doings and suffering, of a thing, whereby the thing may, as a logical possibility, be thought of as changing on its own independently of its relational attributes, then Forms will not have histories. Forms could not have histories in this sense and also fall outside the class of the alterable.[39] When Plato says that we may correctly only say "is" of a Form (37e6), he is referring to the *status* of the Form.[40] He is not trying haplessly to invent a new tense, the timeless present, but is referring to the complete existence which is a mark of measures or standards and which Forms possess as being standards unqualifiedly (*Republic* VI, 504c1–3). It is apposite for Plato to mention the status of the Ideas at this point in the discussion of eternity (*Timaeus* 37e6); for the sort of eternity which the Forms possess is one of the standard-establishing properties of Forms and so is a component in defining the sort of ontological status which Forms have.[41]

IV. ETERNITY AND COGNITION

It is generally recognized that Plato posits his Forms as eternal objects for an epistemological end. On my account, Plato assigns eternity to the Ideas as a standard-establishing property which enables the Ideas to have the functional properties that they have in virtue of serving as standards. I wish to criticize two other, widely influential views of the relation between eternity and cognition in Plato, the views of Hintikka and Owen. I will then end by suggesting that the eternity of the Platonic Ideas is not problem free.

A. Hintikka's View Considered

Hintikka supposes that Plato, following a general trend in Greek thought, viewed informative sentences as being temporally indefinite,

---

[39] Owen wrongly supposes that in the *Timaeus* having a history, even in the weak sense, counts as undergoing a change sufficient to rule out that which changes in this way as eternal, and so further Owen wrongly claims that the *Timaeus* denies that Forms can have histories. Owen (2) 335, 339.

[40] So correctly Cherniss (8) 344–45.

[41] Owen supposes (without argument) that the only intelligible way to speak of a thing as existing is to speak of it as being in time and therefore he holds that it is a contradiction to claim, as Plato does in the *Timaeus*, that Forms exist but are entirely outside of time. Owen (2) 335. Owen supposes that Plato remedies this plight of his Forms in the *Sophist*, where Owen supposes that Plato finally comes around to the view that Forms after all do have histories and so exist in time, as the result of their capacity to come to be known. Owen (2) 336–39.

that is, as having truth values that will depend upon when the sentence is thought of as being uttered.[42] The truth of a temporally indefinite sentence may vary depending upon the moment when it is thought that the speaker could explicitly add the term "now" to a sentence without otherwise changing the sentence. Thus, the truth value of the temporally indefinite sentence "it is raining" may very well differ depending upon whether it is supposed that the statement is being made yesterday or today. Such sentences are not ones "to which we assent or from which we dissent once and for all."[43]

Hintikka supposes that this tendency to view sentences as temporally indefinite is a consequence of the general Greek world-view, which looked at everything from the perspective of the present.[44] Be that as it may, if we add to this tendency, a further (more obviously Platonic sounding) tendency to view knowing as a form of direct acquaintance, which on Hintikka's understanding significantly registers in eyewitness reports of the form "I have seen that P is the case," then listeners to such reports will be assured of the truth of the temporally indefinite claims reported only if the objects reported upon never have occasion to change. These two general tendencies of Greek thought then jointly, for Hintikka, explain why Greek philosophers supposed that the objects of thought were eternal (or at least everlasting).

If I have been correct in suggesting that Plato is viewing time fundamentally in terms of measurement and that the eternity of the Platonic objects of knowledge is a standard-establishing property, then Hintikka's account completely misses both the motivation for and nature of the eternity of Platonic Forms. In particular the assimilation of Platonic statements to a temporally indefinite mode is unfounded.

In establishing a clock as a standard of measure, Plato is clearly not thinking of statements about the phenomenal realm as being temporally

---

[42] Hintikka 8–9.

[43] Hintikka 2.

[44] Hintikka 11–14. Hintikka supposes that Plato is committed to viewing sentences as temporally indefinite specifically as the result of his universal flux doctrine. Allegedly, universal change makes true statements impossible "only if the statements in question primarily pertain to the moment of their utterance and to other moments of time only insofar as things remain constant." Hintikka 8. However, this allegation misrepresents the flux doctrine in that it (commonsensically enough) presumes that at any moment a thing may be accurately identified with respect to kind. Plato, to the contrary, thinks that true statements are impossible in a world in flux because he holds that an object while changing possesses contradictory properties: while changing from being F to not-F, the thing changing is both F and not-F, and so cannot be accurately identified as to kind. Therefore, if all things are at all moments changing, there will be no true statements. For this analysis of the flux doctrine, see chapter 3.

indefinite. For on Plato's understanding of the use of standards, it is a necessary condition for a clock to serve as a measure that changes and states of affairs can be assigned precise dates. This is a pre-condition for measurements being made of them (see section IIC above). Since clock-measured time is part of the backdrop of Plato's metaphysics of the phenomenal realm as it exists for us, judgments made of that realm need not be temporally indefinite; rather, in virtue of implicitly or explicitly having a reference to a determinate date attached to them, they will be true or false independently of the moment when it is supposed that they are uttered, as in "In Oakland it is foggy on the morning of VJ Day." Such a statement will be either true or false (whichever) for all time.

Even statements which express merely relative temporal comparisons and measures are not temporally indefinite. Raw statements of temporal comparison, even without precise dates assigned to their relata, will be true or false for all time. The "is" in the sentence "The Greek victory at Salamis is prior to/before the English defeat at Hastings" is a tenseless usage. The truth of this statement does not depend upon the "is" picking out any particular present moment as the time of its (implied) articulation, even though it may be assessed as being true from the perspective of any moment thought of as the present moment.

Plato then in casting time in terms of measurement raises the average cognizer out of the idiosyncrasies of his own peculiar conditions, including temporal conditions: the truths of his claims are not contingent upon peculiarities of his own history. In this regard Plato's views of time are part and parcel with his general attacks on the Sophists and the purely personal relativism expressed by the ever popular sentiment "But, it's true for me." If in consequence, we must say that Plato's mind is not part of the Greek mind, this is not an accusation at which I suppose Plato would blush.[45]

B. Owen's View Considered

Owen takes Plato's assigning of eternity to the objects of knowledge to be based on a confusion of categories on Plato's part. Plato, Owen claims,

---

[45] It is also likely that Hintikka is mistaken in assessing the motivation and import of Plato's viewing knowing as a sort of direct acquaintance. What is particularly relevant in the knowing-as-seeing analogy for Plato is not that seeing is a necessary prerequisite for popularly intelligible *reportage,* but that no one other than the one seeing can do his seeing for him; reportage assumes that the senses of terms used in reports are already known. But, if knowing is to a large extent for Plato coming to be aware of concepts and meanings, and if knowing is like seeing, then while your 'reports' subsequent to your 'seeing' may have some heuristic value to my understanding, they are not going to be independently in-

has fallen into the confusion of mindlessly taking predicates which properly apply only to statements and applying them to the objects about which the statements are made, presumably in an attempt to explain why the statements have the predicates they do. In the case at hand, Plato allegedly assumes that since some statements are tenseless, they must be about objects existing in a timeless condition and in this way Plato allegedly generates eternal objects in the same illegitimate fashion that some philosophers manufacture necessary beings out of the logical necessity which some sentences possess.[46] Owen supposes that this logical confusion leads Plato in the *Timaeus* to view all and only statements about Forms as tenseless and all and only statements about the phenomenal realm as tensed. Owen holds that this proprietary grip of tenses upon objects is broken only in the *Sophist,* where Forms are, allegedly for the first time, allowed to have histories. However, we have already seen that, on the one hand, in the *Timaeus* the sort of eternity which Forms have does not bar them from having the sort of "weak" histories which Owen has in mind and on the other hand, since the phenomenal realm provides fit subjects for measurement, there will be tenseless sentences even about transient objects (say, skirmishes at Salamis and Hastings). So there is no confusion of categories in Plato's mind here; he has not mindlessly transferred predicates from statements to things.

What is true for Plato is that all statements that *only* have Forms as their objects will be tenseless, statements like "Three is an odd number." Moreover, once measure is admitted into the phenomenal realm, tensed (but temporally indefinite) sentences which have variable truth values and which make claims about the phenomenal realm, sentences like "the number of congressmen in jail is three," can be converted into "more stable" sentences. By reference to the temporal standard, they can be recast as sentences with fixed truth values, like "the number of congressmen in jail on VJ Day is three," a statement to which we assent or from which we dissent once and for all. It turns out then, as I have suggested from the start, that the presence of measures and standards in the phenomenal realm makes our cognition of that realm more like our cognition of the Ideal realm. It is for such an end that Platonic time is created.

C. A Problem of Causal Inertness

The eternity of Plato's Forms does present a problem for his epistemology (and his metaphysics generally). The type of problem will be famil-

---

telligible to me, prior to my own 'seeings'. Hintikka's account will be all the more mistaken if, as I think is the case, there is an element of each Form which is not capable of being cast into discursive modes of representation.

[46] Owen (2) 336.

iar from recent debates over the possible intelligibility of the abstract entities hypothesized as useful or necessary by many philosophers of mathematics. The problem, stated generally, is that if entities are of such a status as to fail to have causal efficacy, then they will be incapable of entering causal chains of events by which it would be possible to come to know them. The problem may be called the causal inertness problem, and for Plato it takes the following form. If, in order to serve as standards, the Forms have a status which requires their not falling within the category of things which either alter or do not alter and so *a fortiori* do not alter in any way, then they seem to be entirely removed from the realm of causal interactions, and therefore cannot, after all, serve as objects of knowledge.

The reasonableness of a demand that the objects of knowledge form at least part of the effective causes of our cognitive states in order for those states to count as knowledge will be especially compelling on models of knowing on which knowing is a form of direct acquaintance.[47] For on a direct acquaintance model, whatever justifications one has for one's cognitive states have to be built right into the mechanisms which form those states. The reliability or infallibility of the mechanism is not itself subject to doubt and scrutiny on this model in the way it is on a model which views knowledge as *justified* true belief. But clearly we can have cognitive states which are based on direct acquaintances and which are true and yet which we would rule out as knowledge since the object of the acquaintance and the object of the true cognitive state are merely accidentally related. Some of our direct acquaintances in dreams and of after-images will produce such states. Thus, if someone happens to place a green patch where I am seeing a green after-image and if I believe and claim that there is a green patch "there" (pointing to the patch), we will still not count my belief as knowledge even though it is a true belief and even though the belief is based on a direct acquaintance. But, in consequence of the direct acquaintance model, I cannot rule this out as a case of knowing on the ground that the belief is unjustified for some reason (that, for instance, viewing after-images is an extremely unreliable procedure for justifying beliefs). On the direct acquaintance model, I can rule it out and can *only* rule it out on the ground that the object of my true belief failed to cause my state of cognition. My true belief about the green patch was not based on any evidence that involved the patch.

---

[47] A causal condition is generally thought to be required on the now "traditional" model of knowledge as justified true belief, in order to avoid the Gettier problem, to wit: a person, S, justifiably believes a false proposition *q*, infers from this that '*p* or *q*', where the disjunctive proposition '*p* or *q*' is true. Since S has no evidence for *p*, S does not know '*p* or *q*', even though S has the justifiably true belief that '*p* or *q*'. See Pappas 12.

If, then, "knowing" or intuiting is going to be a kind of direct acquaintance which is even a possible candidate as a mode of knowing, it will have to be like (clear) vision, and unlike dreaming and having after-images, *in* (at least) *the respect* that the judgments produced by the faculty are (in part) caused by the objects about which the judgments are made. If, as seems to be the case, Forms are causally inert, they will fail to provide the requisite causal condition for their being known. They will fail to provide any evidence for our (possibly) true beliefs about them.

In the *Republic* Plato tries unsuccessfully to get around this problem of causal inertness with his analogical comparison of the Form of the good to the sun. The Form of the good is supposed (among other things) to make the other Ideas knowable in the way that the sun makes phenomenal objects visible (508b–509b, 517c). Presumably the relevant part of the sun analogy here is that the sun emits light which is encoded with information from the objects off of which it reflects; this information is then passed on by the reflected light to the observer of the objects. Knowing is, thus, viewed as a sort of irradiation of the mind by the Ideas (*Republic* 490b, 508d, 517c; *Phaedrus* 250c). But if the Forms are utterly immutable, the analogy breaks down at two points. First, there can be no interaction between the 'sunlight' and the Forms which it illuminates and so no encoding of the reflected 'light'. Second, there will be no emanation of 'light' from the 'sun' to begin with (note especially παρασχομένη, 517c4).

The problem of causal inertness is not limited to Plato's epistemology. For several claims of his metaphysics and psychology also seem to require causal efficacy of Forms. The soul, for instance, is repeatedly claimed to have a (contingent) immortality, as the result of being constantly nourished and nurtured by the Ideas (*Phaedo* 84a–b; *Republic* 490b, 611e; *Phaedrus* 247d–e, 249c). Even with allowances made for mythical speech, it is hard to know how to construe these claims if they do not entail causal efficacy on the part of Forms.

Moreover, even if the *Timaeus* myth is read non-literally, it is clear that Plato supposes that the Forms and the receptacle of space are jointly sufficient for the appearance of (some aspects of) the phenomenal world in the receptacle. This will hardly be possible if the Ideas are not thought of in some way as self-illuminating and as causing their images to be cast upon the mirror of space. Attempts to explain the role of Forms in Platonic participation as simply being causes in the way Aristotelian causes in general are causes, without, that is, having the slightest taint of causal efficacy, have not been particularly successful[48] and in any case only exacerbate the problem of causal inertness within Plato's epistemology.

One can then accept the sense of eternity which attaches to Forms only if one concomitantly accepts a certain degree of paradox in the way in

[48] See Annas (3) 311–26.

which they inform the world—and *Phaedo* 100d–e may in fact be signaling that Plato himself did accept such a stance. Whether one is willing to admit some degree of paradox here depends of course on the degree to which one otherwise finds the theory of Forms, as a whole, an exclusively attractive solution to the problems of knowledge.[49]

---

49 Those who have found Plato's thought generally coherent have also straightforwardly admitted that it does contain paradoxical elements. See Shorey 30, Cherniss (7) 309 ff.

PART TWO

THE EFFECTS OF NECESSITY

Timaeus 48a–69a

THREE

Image, Flux, and Space in the *Timaeus**

Heraclitus wrote that you cannot step into the same river twice for the waters are ever flowing new upon you (B49a, B12, B91, A6). We are told by Aristotle (*Metaphysics* 1010a11–14) that Cratylus, a follower of Heraclitus, revised this slogan to read "one cannot step into the same river even once," and then stopped talking altogether. Cratylus believed that nothing whatsoever may be predicated of the phenomena in flux.

We are also told by Aristotle that Plato was a follower of Cratylus and adopted Cratylus' views into his metaphysics (987a30–b1). In this chapter, I will discuss to what degree this is true. The chapter has two sections. The first gives a new analysis of the tortuous and much debated passage of the *Timaeus* in which Plato discusses the possibility of making predications of the phenomena in flux and in the course of which he formally introduces his conception of space. The passage is *Timaeus* 49b–50b.[1] The second section is a general discussion of Plato's conception of space.

I

The second part of the *Timaeus* (47e ff.) where Plato discusses the nature of the physical world begins by assuming that some problem confronts all who try to discuss the nature and generation of earth, air, fire, and water. The problem is to say what each of these severally is (πῦρ ὅτι ποτέ ἐστιν καὶ ἕκαστον αὐτῶν, 48b6–7). In this way, the second half of

---

* Reprinted from *Phoenix* 34 (1980) by permission of the editor.

[1] The passage has proved endlessly fascinating to critics. See Cherniss (9), Cherry, Cornford (2) 178–85, Gill, Gulley, Johansen 117–36, Lee (1), (2), (3), Mills, Reed (1), Silverman (1), (2) 246–84, Zeyl (1), (2) lvi–lxvi. Since the appearance of *The Platonic Cosmology*, my interpretation has been attacked by Gill, Gill's by Silverman, and Silverman's by Zeyl. I do not believe that the criticisms that Zeyl launches against Silverman's (and Cherniss') interpretation have force against mine. Zeyl (2) lix–lxvi. Indeed Zeyl appears to make a silent nod toward my interpretation in an addendum to his own interpretation. Zeyl lxv–lxvi, discussed below.

the *Timaeus* begins like many of the Socratic dialogues by asking the "what is it?" question. The question, which is repeated at the start of the gold analogy (50b1), is an internal question which asks us to *identify* various phenomenal kinds (in the first case, earth, water, etc. and in the gold analogy, various geometric figures).[2] The "what is it?" question is not here an external question which asks after the ontological status of an object. Plato is here interested in how earth can be distinguished from water, rather than whether it is a substance, quasi-substance, quality, quantity, a form, a matter, an Idea, etc. In order to answer the question of how one identifies earth, water, etc., we are told that a third sort of entity must be added to the furniture of the universe. For the first half of the dialogue, two sorts of entities were sufficient. The one is the Ideas, which are spoken of as paradigms or models, and as being intelligible and always unchangingly what is; the second is the phenomena, which are copies of these models, are visible, and though in the process of becoming, never are (48e3–49a1). This "becoming" (cf. 52d3) refers to the precosmic or acosmic flux of the phenomena described at 30a and 58b–c. The third thing which must now be added to the ontology is called the receptacle of all becoming and is later identified with space (52a8). Why this third thing is required we are not yet told; that requires elaborating the earlier difficulty of identifying phenomenal kinds. The first stage of the elaboration is contained in the following sentence: "It is difficult to say with respect to each of these kinds (earth, water, etc.) what type (ὁποῖον) we really ought to call water rather than fire, and what type (ὁποῖον) ought really to be spoken of as a particular type of thing rather than as all types taken together or severally, so that we really use language in a *somewhat* (τινί) reliable and stable way" (49b2–5). The problem, it seems, is not that various phenomenal types shade off into each other, in such a way that, say, we might be confused as to whether a patch of rust should be called orange or brown. Rather, it is that all phenomenal types seem to be interchangeable so that all the phenomena are equally well called by any and all names of types of phenomena.

Why the phenomena are in this strait has not been fully stated. But with the statement that we are eventually to end up with language that is *somewhat* stable and certain (49b5), we are already given a hint of the solution to the problem. For this phrase harks back to the principles which open Timaeus' discourse at 29b–c. There it is said that language

[2] Throughout the chapter by "identify" I mean to pick out an individual as the kind or type of individual it is, as in the sentence, "the chemistry student correctly identified the samples as zinc." I do not mean "identify" in the sense of picking out an individual as the individual it is, as in the sentence, "the victim identified the culprit in the police line up." Zeyl uses "identify" in this latter sense. Zeyl (1) 128, 130, *et passim*. Where Zeyl uses the term "identification," I use the term "individuation."

will be as certain as its objective referents are stable; language about the immutable Ideas will be absolutely certain, while language about becoming will be certain to the extent that becoming is a *likeness* of the Ideas (29c1–2). The hint, then, is that the phenomena may be identified insofar as they are likeness or images of the Ideas. On the other hand, the first stage of the elaboration of the problem of identifying the phenomena, in saying that a phenomenal object in and of itself can have any and all predications applied to it, more or less accuses the phenomena of being referents for self-contradictory propositions. Plato makes the corresponding ontological claim in the *Phaedo* where he claims of the phenomena, just as phenomena, that they are "never the same as themselves" (*Phaedo* 80b4–5).

To assert this logical and ontological contradictoriness, though, is not to assert directly that the phenomena in no way exist, but rather, I suggest, is to assert only that the phenomena are in flux (cf. *Phaedo* 80b4–5 with *Philebus* 59b1).[3] This is how Aristotle interprets the predicational schemas of the Heracliteans (*Metaphysics* 1010a6–37) and contradictory propositions seem to be entailed by Plato's analysis of change in his discussion of "the suddenly" in the *Parmenides*.[4] Note also the description at *Timaeus* 57c3–4 of things that are changing as things which are becoming different from themselves.

That the first stage of the elaboration of the problem of identifying phenomena suggests that the phenomena are subject to contradictory predications should leave no surprise, then, that the final elaboration of the problem is the assertion that the phenomena are in flux. The reason the names of all types fit each phase of the phenomena is that the phenomena are in a universal flux, in which every type is constantly transmuting into all other types, as described at 49b7–c7, where water is traced through a cycle of transformations which turns water into all the other three types of primary bodies.

The difficulty of answering the original "what is it?" question, then, is the difficulty of being able to *identify* or state the kind (ποῖον) of phenomenal objects, given that they are in flux and transmute into each other.[5]

---

[3] Some critics wish to claim that for Plato, though the phenomena are in some sense disparate, they are in no way (self-)contradictory. See for instance, Irwin (2). *Phaedo* 80b4–5 and *Timaeus* 49b2–5, however, would seem to undermine this claim.

[4] *Parmenides* 155e–157b, and cf. 156e7–8 for the full generality of the doctrine. Here it is claimed that when a thing is changing from having a property to having its opposite, it passes through what is called "the suddenly" in which it has neither the property nor its opposite (156d). This is logically equivalent to a denial of the law of non-contradiction.

[5] Johansen correctly diagnoses the difficulty that the passage is asserting. Johansen 120.

This difficulty is finally fully stated at 49c7–d3. The sentence fuses the problem of identification with the assertion that the phenomena are in flux: "Accordingly since each of these (earth, fire, etc.) never appears the same [i.e., is in flux], which of these types [ποῖον] could one steadfastly affirm without embarrassment as being this particular type and not another? It is not possible." The rest of the passage which gives the answer to this problem of identification runs as follows:

> Rather, the safest way by far to speak concerning these is by proceeding in the following manner. What we see, since it always is becoming at different times and in different ways—for example, fire—we should not say on each occasion that this is fire[6] but that something of a certain kind [τὸ τοιοῦτον] is fire, and on each occasion say not that this is water, but that something of a certain sort [τὸ τοιοῦτον] is water, nor should we call <this>, as though it had some stability, anything else [say, "air" or "earth"] which we suppose reveals something when we point and use the terms "this" and "that" [as in "this is fire"]. For it flees and does not abide the assertion of "this" or "that" or any assertion that indicts them as being stable. But instead of calling them [earth, fire, etc.] each of these things ["this," "that," etc.], we should thus call these [the fleeting phases of the phenomenal flux] "a certain type" [τὸ τοιοῦτον], which always recurs similarly in each and all cases; for instance, fire should be called "that which is on each occasion of a certain type" [τὸ . . . τοιοῦτον], and so with everything that goes through transformations. But, that *in which* each of these appears and again out of which they pass away [i.e., the receptacle or space], this alone by contrast ought to be designated by the use of the terms "this" and "that," but what is of any kind whatsoever, hot or white or any of the contraries and all that consists of these, this [i.e., space] should not be called any of these ["hot," "white," etc.].[7]

At 50c4–5 the referents of "a certain sort" (τοιοῦτον), which are said to enter into and leave the receptacle, are explicitly equated with images of the Ideas.[8] The passage, then, is to be interpreted as follows.

The passage reconfirms the view that the phenomena *as in flux* cannot be identified as to type. Nevertheless, the passage maintains that the

---

[6] With Cherniss and Mills, I take the presence of πῦρ (d6) as decisive against taking τοῦτο and τὸ τοιοῦτον as the secondary objects of προσαγορεύειν (d6). Cherniss (9) 349, Mills 154n14. Πῦρ at d6, by position, is obviously not resumptive, as Cornford and Zeyl seem to take it. Cornford (2) 179, and Zeyl (1) 132.

[7] On the immediately ensuing gold analogy, see chapter 4.

[8] While correctly diagnosing the problem the contested passage states, Johansen gives only passing mention of the fact that the things that come into and pass out of the receptacle are images and so can only pass in and out of the receptacle in the way that images pass into and out of a medium. More generally, Johansen fails to incorporate the contents of the receptacle as images into his interpretation of the metaphysical role(s) of the receptacle. Johansen, especially 121.

phenomena can be identified and are subject to the predication τοιοῦτον, "of a certain sort."[9] It is as recurring rather than as being in flux that a type of phenomena may be identified. What recurs in the flux of phenomena are images (50c4–5). We can tell that what recurs is the same recurring image by referring it to the original of which it is an image. The phenomena, then, have a double aspect. On the one hand, they are in flux; on the other hand, they are images of Ideas. Insofar as the phenomena are in flux, nothing whatsoever may be said of them. But insofar as they are images of Ideas, they may be identified according to kind.

The introduction of the phenomena as images solves the problem of identifying the phenomena. This solution gives the formal reason for the introduction of the receptacle. The third thing is needed as a medium in which the phenomena, as images, may appear. This need was not felt, though, until the problem of identifying the phenomena had been clarified (49b–c). That the phenomena are in flux required some means other than direct appeal to the phenomena themselves in order to identify them. The third thing having the nature and power (49a4–5) of a receptacle and medium is a requisite for fulfilling this need, since it makes it possible for the phenomena to appear as images and so makes possible the identification of the phenomena insofar as they are images of Ideas.

Further by drawing into doubt the ontological status of the phenomena as being in flux (especially 50b3) and by contrasting the phenomena as in flux with the phenomena as images, our passage leads the way to the assertion that the shifting phenomena *as images* in the receptacle "cling in some way to existence on pain of being nothing at all" (52c3–5).[10] Thus is resolved the problem, left over from the *Republic*, of how becoming holds a middle ground between being and non-being: that the phenomena are reflections of Ideas saves them from the non-existence and complete unintelligibility which threatens them as a result of their being in flux.[11]

Though my interpretation of the passage is simple, it has eluded Platonic scholarship. Some critics (notably Zeyl) have construed the problem with which the passage deals as a problem of how the phenomena are individuated rather than identified and claim that in the passage sensible phenomena are not in themselves substantial, but are made sub-

---

9 The claim that the phenomena are τοιοῦτον ("of a certain sort") answers the question ποῖον; ("of what sort?") (49d1). Τοιοῦτον is simply the demonstrative pronoun correlative with the interrogative ποῖον.

10 Johansen correctly sees that the phenomena's being in flux draws their existence into doubt, but he does not see their status as images as the answer to this doubt. Johansen 118–21.

11 For the relation of *Timaeus* 49b–50b to the *Cratylus* and *Theaetetus*, see chapter 4. See also Reed (1).

stantial by their relation to the receptacle, which individuates them and serves as a substrate for their changes. For these critics, τοιοῦτον does not distinguish an aspect of the phenomena as images from their aspect as in flux, but rather applies even of the phenomena as in flux and denotes a form-like quality of the phenomena (in Aristotle's sense of form). For these critics the whole passage looks forward to Aristotle's form-matter distinction.[12] The formal reason for the introduction of space, on Zeyl's account, is to serve as a principle of individuation and as a substrate for change.[13]

Three other critics (Cherniss, Lee, and Mills) form a group which claims the passage has nothing to do with form-like properties; each of these critics, though, takes τοιοῦτον as picking out a different sort of entity. Cherniss takes τοιοῦτον correctly as referring to images of Ideas, but wishes to separate the images from the phenomena in flux, by equating images rather with what he calls "self-identical characteristics," which, he claims, cause the phenomena to come into existence by their entrances into space.[14] The chief objection to this view is that there are at least four

---

[12] Zeyl (1) 146–48, (2) lxii–lxiv; so too, Johansen 132–36, Algra, chapter 3, and Miller, who gives the receptacle a very narrow interpretation as a material principle hypothesized just to provide a material base for the triangles that make up Plato's geometrized primary bodies (earth, water, etc.). Sort of as an afterthought, Zeyl admits that Plato's treatment of phenomena as images offers a secondary justification for the existence of a medium in which they must appear if they are to appear at all. Zeyl (2) lxv–lxvi.

[13] Zeyl's understanding of the problem with which the passage deals and its solution are the result of (1) his translating, like Cornford, ὁποῖον (49b3, 4) and ποῖον(49d1) quantitatively rather than qualitatively and so failing also to see that ταὐτὸν τοῦτο (49c1–2), τούτων (49c7), τῶν αὐτῶν (49d1), and αὐτῶν ἕν(50a7) refer to types and not to individuals, (2) failing to see that the problem raised by the flux of the phenomena is a problem of identifying them rather than finding a permanent substrate for them, i.e., failing to see that 49a6–b2 refers back to 48b5–8, and (3) failing, like Lee, to see (a) that the stability of discourse made possible by our passage is a qualified one and (b) that 49b5 refers back to 29b–c. Zeyl (1), Lee (2) 22.

Zeyl, however, is right, as against Cornford, in seeing that τοιοῦτον can denote properties in any category (substance, quality, quantity, relation, etc.) and is not restricted just to the category of quality, as Cornford would have it. Zeyl (1) 147. Cornford confuses two senses of substance in his analysis of our passage. Cornford wishes to claim that our passage in calling phenomena τοιοῦτον treats phenomena as non-substantial, where substance is understood as a composite of form and matter, but he then unwarrantedly assumes from this claim the further claim that τοιοῦτον cannot denote substances, where substances are understood as essences or natural kinds as opposed to accidents. Cornford (2) 178–81, especially 181.

[14] Cherniss (9) 361–63. Cherniss' view is defended by Silverman (1).

places in the dialogue, three in the immediate context, where the phenomena as images and the phenomena as in flux are taken as extensionally equivalent (29c1–3, 48e6–49a1, 50d1–2, 52a4–7).[15]

Lee, though correctly seeing that images and that which is in flux are co-extensive, nevertheless claims, like Cherniss, that τοιοῦτον denotes self-identical characteristics, which hover in an ontological twilight somewhere between the Ideas and the phenomena, but which are not images of Ideas.[16] This interpretation, however, collides with the explicit equation of that which is denoted by τοιοῦτον and the images of Ideas at 50c4–5.

Mills goes so far as to suggest that the referents of τοιοῦτον are the Ideas, despite the fact that the Ideas are explicitly said not to enter into anything whatsoever (52a3) and in particular not to enter into the receptacle (52c6–d1).[17] Indeed, this property of the Forms is the very reason given at 52a–b for differentiating Forms *from* their images.

The convolutions of the Platonic metaphysics wrought by Cherniss, Lee, and Mills are motivated by an attempt to get around asserting τοιοῦτον ("of a certain kind") of the phenomena. For they take such asserting to be denied in the sentence of my translation which starts "what we see . . ." (49d4–6). Here I take the participial construction *causally* rather than purely *descriptively*, rendering it "*since* it always is becoming . . ." and taking it to mean "*insofar as* it always is becoming." This construction and interpretation is justified by the fact that the phrase is a summary of the preceding description of the phenomenal flux (49b7–c7), which raised the difficulty of picking out a phenomenal object as being of a certain kind. I suggest then that my simple interpretation of the passage holds.

The problems (insurmountable in my view) which beset the interpretations of Zeyl, Cherniss, Lee, and Mills are dissolved if we read 49b–50b along the lines I have suggested. The passage distinguishes but does not separate the phenomena viewed as being in universal flux from the phe-

---

[15] Despite this overwhelming evidence, Cherniss claims that 50c3–4 supports his view by its implying that the transient phenomena are apparent alterations of the receptacle induced by copies of Forms entering it. Cherniss (9) 362. Rather this sentence, which in context is offered as a reason (γάρ, c2) why the receptacle is characterless and unchanging, simply warns against inferring any diversity of the receptacle itself from the diversity of the phenomena in it. It is the diversity of the receptacle that is apparent, and, as we are told elsewhere, altogether impossible (50b8–c2, 50d–51a). The first δέ 50c4 is adversative; φαίνεται (c3–4) here means "appears to be *x*, but indeed is not *x*" rather than "is visibly represented by *x*" or "shines forth visibly as *x*."

[16] Lee (1) 366–68, (2) 22–25.

[17] Mills 154.

nomena viewed as images of Ideas. As in flux, the phenomena cannot be identified according to kind. The mutability of the phenomena draws into doubt their intelligibility and even their very existence. On the other hand, as images of Ideas, the phenomena are subject to the predication τοιοῦτον, are saved from utter non-existence, and can be identified according to kind. As images of stable being, the phenomena are subjects for discourse which is stable to the degree that its objects are likenesses of the Ideas.

II

I now wish to spell out the consequences that our newly interpreted passage has for Plato's overall conception of the third thing or space. I will do this in two parts, first by showing what light our passage throws on the various metaphors which Plato uses to describe the third thing or space and second by discussing how the passage helps us determine whether or not Plato's conception of space operates philosophically like Aristotle's conception of matter.

Plato uses two groups of metaphors to describe the third thing. One group of metaphors treats the third thing as a container and suggests a "bucket" theory of space. The other group of metaphors treats the third thing as a mirror or more accurately as a medium or field for receiving images.

The third thing is a container insofar as it is called a receptacle or reservoir (49a6) and space (52a8) and is compared to a winnowing-basket (52e6) and to a place for sitting (ἕδραν, 52b1). Philosophically such metaphors of the third thing as a container serve to distinguish Plato's conception of space from conceptions which take space as equivalent to material extension. Such conceptions in general (1) make space that out of which phenomenal objects are made and (2) make space ontologically dependent on the existence of phenomenal objects. By using the container metaphor Plato wishes to deny both (1) and (2). On the metaphor of a container, the third thing with its contents is rather like an urn full of potsherds. It makes sense on this metaphor to say that one can place objects in the receptacle and again take them out keeping each intact, unaltered, and numerically one. That this is so, however, I suggest, is a non-relevant consequence of the container metaphor and is not intended to have metaphysical implications.

For, on the other hand, Plato speaks of the third thing as though it were a mirror (49e7–8 with 50c5). He does not specifically use the term "mirror" of the third thing for two reasons, both related to the elaborate account of mirrors which he appends to his recent discussion of the mechanism of vision (46a–c). First, for Plato the appearance of images in mirrors entails the existence of a perceiver (46a–c, cf. *Sophist* 266c) and he does not wish to imply that the relation between the Ideas and the receptacle entails the presence of a perceiver (see 61c–d). Plato may be a phenomenalist with regard to perception, but he is not a subjective idealist. Second, Plato's account of mirrors emphasizes the way mirrors "flip"—right-left reverse—the images that appear in them (46a–c), while the discussion of space repeatedly emphasizes that space intrudes no character into that which it receives (50d–51a). At *Timaeus* 71b–d, Plato views the liver as a medium for receiving images. The images it receives are dreams, nightmares, by which reason terrorizes the appetitive part of the soul. This medium for receiving images can be—and is—explicitly compared to a mirror (71b4), since, unlike the receptacle of space, it 1) is intentionally designed, like a crinkled fun-house mirror, to distort images, and 2) has a perceiver, the appetitive part of the soul.[18]

Plato does explicitly use metaphors which portray the third thing as a medium, as opposed to a container. Such are the comparisons of the third thing to a scent base (50e) and the description of the third thing as ἐκμαγεῖον, a soft substance capable of receiving impressions (50c2, cf. 50e8–9). And viewed functionally rather than descriptively, the third thing as compared to a mother (50d3) should be ranked as a medium,[19] as should perhaps also the metaphor of the third thing as a nurse (49a6). Now philosophically the metaphor of space as a mirror or medium serves to show the ontological dependence of the phenomena on both the Ideas and the third thing and suggests that only the third thing and not its contents is stable and so subject to being called "this." The metaphor also serves to distinguish the third thing from those conceptions, including Aristotle's, which view space as that which surrounds on the outside the extension of phenomenal objects. These philosophical functions do not hold for the metaphor of the third thing as a container, where objects in it are surrounded by it, are stable, and are ontologically independent of it, like marbles in a goldfish bowl. So Plato while setting forth the philosophical functions of the third thing uses metaphors which, if read literally, would, when taken together, assign incompatible properties to the contents of the receptacle. Which set of properties should we choose?

---

[18] Kung has denied that Plato intends space to be thought of as a mirror. Her analysis is marred by a *presumption* that Aristotle must be right when he views Plato's receptacle as a forerunner of his own material cause. And she fails to take into account reasons why Plato would leave the mirror metaphor tacit.

[19] See Cornford (2) 187.

It is clear from our passage *Timaeus* 49b–50b that it is the metaphors of space as a medium for receiving images which are meant to be taken at face value and that the reading of the other metaphors must be brought into accord with the conception of space as a medium. And indeed this can be done.

For Plato is not committed to the contents of the third thing being like the contents of a container, since the philosophical points Plato wishes to make about the third thing by using the container metaphor hold also of the third thing viewed as a mirror, even though the mirror metaphor does not make the points as perspicuously as does the container metaphor. A mirror is not equivalent to the extension of objects in it, nor is it ontologically dependent on its contents, nor is it a material constituent of its contents. We need not then accuse Plato of any philosophical confusion about the status of the third thing and the properties which the phenomena have as phenomena in it. Whereas Plato is committed to seeing the receptacle's contents as being like the contents of a mirror (each mention of images present in something makes such a commitment, especially 52c3–5), Plato is not committed to viewing the contents of the third thing as the contents of a container. Nor does he make appeals to the contents of the third thing viewed as contents of a container.[20]

When, then, Plato says that the contents of the receptacle appear in and then vanish out of the receptacle (49e7–8), he does not mean (and cannot consistently maintain) that the contents of the receptacle can be placed in and taken from it, each remaining intact, unaltered, and numerically one, like marbles in a goldfish bowl, though Cherniss' self-identical characteristics would commit Plato to this contradictory view. Rather the contents appear in and vanish from the receptacle as images in a medium. "To vanish" here, then, means not to go elsewhere, but simply means that the image's original stops being projected into space.[21]

We can clarify Plato's conception of space even further if we compare it to Aristotle's conception of matter. Philosophers and critics of many persuasions have from Aristotle to the present tended to identify the two, taking Plato's conception of space as a failed formulation of Aristotle's conception of matter.[22] This view is maintained I suggest largely by taking various phrases which Plato uses to *describe* the receptacle and

---

[20] For a discussion of Plato's comparison of the receptacle to a winnowing basket (52e6–7) , see chapter 6.

[21] Johansen misses this crucial point about coming into and going out of the receptacle. Johansen 121.

[22] This view has been especially strong in the years lapping into the twenty-first century. Kung, Zeyl (1) 146–48, (2) lxii–lxiv, Johansen 132–36, Algra, Miller.

wedding these to Aristotle's *descriptions* of matter without seeing what philosophical roles the two entities play in each philosopher's system. By viewing Plato's space as a permanent, characterless substrate which is known only by some non-standard mode of cognition (cf. 51a–b), these critics make Plato's space look much like Aristotle's matter, which is permanent, lacks internal properties, and is known only by analogy. However, if we distinguish the philosophical *functions* of the two sorts of entities in each philosopher's system, we will see that they have little in common. There are five basic functions which matter serves (or can be construed as serving) in Aristotle's metaphysics: it serves (1) as a material cause or that out of which phenomenal objects are (made), (2) as a principle of individuation, (3) as a subject of predication, (4) as a substrate for change, and (5) by inference, as a principle of existence. I suggest that only the last function is held in common between Aristotle's matter and Plato's space.

(1) It is largely by pressing the comparison of space to gold in the *Timaeus* that critics have been able to construe space as a material constituent of phenomenal objects or that out of which phenomenal objects are made[23] and so claim that Plato, like Aristotle, is viewing phenomenal objects very much like substances composed of form or shape and matter. I suggest though that Plato's description of the gold as that out of which shapes are formed (50a6) is an exigency of the metaphor and is not the relevant aspect of the gold that is being compared to space. Gold is chosen in the *Timaeus*, I suggest, not to typify matter entering into a substance as the constituent which offers stability to form, as the bronze of a sculpture offers stability to the shape of the sculpture, but exactly because of its malleability. Though the gold in itself is permanent, it does not transmit or contribute its permanence to the phenomena. Just look at the sentence in which the gold metaphor arises (50a5–b1): the figures in gold are being ceaselessly remolded into each other and it is this radical impermanence of the phenomena that raises the difficulty of identifying them. This difficulty is resolved by our ability to identify phenomena *as* images. Appeals to the gold do not help us to identify the phenomena or to view the phenomena as stable. Indeed the gold itself, as that which is ceaselessly molded, is not even being treated here as a medium for images—whether or not these images are treated like Aristotelian forms.

When space *is* viewed as a medium for images (50c), the images are, to borrow an idea from Lee, non-substantial images, that is, images which depend for their continued existence upon the persistence of their origi-

---

[23] See for example, Claghorn chapter 2, Guthrie 253, 262–69, especially 265, Zeyl (1) 142, 147, (2) lxii. Cf. Johansen 122n8, 134.

nals.[24] They are like images in mirrors, or like shadows on cave walls, or like reflections in water, and are unlike statues or paintings of living individuals, where the image may persist, as the result of its material components, even after its original perishes. It is these sorts of images, rather than sculptures and paintings, which Plato uses in the Divided Line and Cave Analogy of the *Republic* to help clarify the ontological status of the phenomenal world. Thus the lowest level of the Divided Line, the segment which typifies the world around us, is said to consist of "images—and by images I mean shadows in the first instance, then the reflections in water and all those on close-packed, smooth, and bright materials," to wit, metal mirrors (*Republic* VI, 509e–510a, Grube). And the lowest level of the Cave, the level that represents the world around us, consists of shadows cast on a wall (*Republic* VII, 515a). The stability of the phenomena, as we already know from *Timaeus* 29b–c, is dependent on the images' relation to the Ideas, rather than on the relation of the images to space. Neither the gold analogy nor any other passage in the *Timaeus* entails viewing space as a material constituent out of which substances are created. Rather space is a medium or field *in which* or *across which* phenomena appear as (non-substantial) images.[25] If Plato had lived into our

---

[24] Lee (1) 353–60. Though Zeyl draws attention to Lee's understanding of the images in the receptacle as *non-substantial* images, when Zeyl finally begrudges the possibility that images might have something to do with Plato's reasons for hypothesizing the existence of the receptacle, all of his examples of images are Aristotelian composites of form and matter, rather than examples of Timaean images that require for their existence both a medium across which they flicker and the persistence of their originals. Zeyl writes: "So the image must be 'in' something, as, for example, an image of Abraham Lincoln must be 'in' marble or bronze or oil on canvass or glossy paper, in order to be an image. There must, then, be a Receptacle for any image *of* Forms to be *in*." Zeyl (2) lxi n129, 42n57, quote from lxv–lxvi. But notice that if a twenty-first century artist carves a marble sculpture of Abraham Lincoln only to see it, at the moment of its completion, blown up in a terrorist attack, we would not say in this circumstance that an image of Lincoln had appeared in and then vanished out of the marble. We *would* say of Lincoln's image that it keeps appearing in and vanishing from mirrors as the President walks down the Hall of Mirrors at Versailles. Indeed we do not even speak of images being in lumps of stuff. Rather when we speak of a sculpture as an image, as in "Thou shalt not make unto thee any graven image" (Exodus 20.4), it is the sculpture as a whole, not an aspect, facet, or a dimension of the sculpture that is the image. Plato's images in the discussion of the receptacle are non-substantial images, not Aristotelian composites of form and matter.

[25] Guthrie claims that Plato uses the expressions ἐν ᾧ and ἐξ οὗ indifferently "in connection with raw material" and cites *Philebus* 59e and *Statesman* 288d as evidence. Guthrie 265n3. In both of these passages, though, Plato is talking of already highly organized materials, which are then, as instruments, put to a final

century, he might very well have chosen, not gold, but a movie screen or television screen as his analogue to a field across which ceaselessly changing non-substantial images may flicker.

(2) For Aristotle one of the main philosophical functions of matter is to serve as a principle of individuation. There are many places where matter is viewed as *a* principle of individuation (*De Caelo* 278a26–27; *Physics* 190b22–25; *Metaphysics* 1074a33–35) and in *Metaphysics* V, he goes so far as to say it is *the* principle of individuation (1016b32–33). Socrates and Coriscus are, as men, formally identical, but because they are made out of different matter, they are numerically distinct. This does not mean that matter for Aristotle consists of pre-packaged parcels of numerically distinct stuff, like Democritean atoms or bare particulars, which individuate qualities because the parcels are in and of themselves individuals.[26] Rather Aristotelian matter individuates, but is not in and of itself individuated. It is like dough to a cookie cutter: the dough individuates the various cookie cut-outs, but does not consist of a group of individuals independently of having been cut into cookies. Now it is clear that Plato's space does not consist of pre-packaged units. He speaks of regions of space, but these regions are posterior to the images which fill space (51b4–6): there is a fire region because fire is extended through part of space. Space does not consist of a grid of cubbyholes, which images of fire may then fill. But I also suggest that space for Plato does not serve as a principle of individuation in Aristotle's sense. Plato certainly never says that space individuates phenomena of the same kind, though space, as a whole, is numerically one. And indeed since space is not that out of which the phenomena are made, it cannot serve as a principle of individuation at least as Aristotle's matter individuates. Consider again the relation of original, (non-substantial) image, and medium. In this relation it is not the medium which individuates the image. There are several ways for the image to be multiplied: the medium may be divided, like a shattered mirror; the original may be reproduced (as shadows of two hands cast on a screen), or the projecting light may be reproduced (one hand

---

use. The referents of both passages are what in the *Timaeus* would be called "accessory causes" (46c) and in any case refer to the receptacle's contents rather than to the receptacle itself. It should be remembered that both the receptacle and its contents, as described in our passage, exist independently of demiurgic crafting. When the Demiurge does start making things, he makes them out of the receptacle's contents, not out of the receptacle itself.

[26] *Physics*, I.7 seems to offer a single exception to this rule: "Now the subject is one numerically, though it is two in form. For it is the man, the gold—the 'matter' generally—that is counted, for it is more of the nature of a this . . . ." (190b23–26, Hardie and Gaye).

may cast several shadows, if there are diverse light sources). But as long as the medium remains intact and numerically one, it does not provide the principle for individuating phenomena of the same kind.

I suggest that Plato does not have a principle of individuation for the phenomena, rather he takes the plurality of instances of the same Form as a *given* (as at 52d5–e2). I suggest that this is also what he intends when, in discussing the sort of existence which the physical world has, he calls phenomenal existence "partible" as opposed to the impartibility of the Ideas (35a). For this partible existence is later called "dispersed, scattered, or strewn existence" (37a6). Images, then, are strewn across space like individual seeds onto a field, or troops scattered across a field in a rout, or sunrays scattered across the land (σκεδαστήν, 37a5; *LSJ*, s.v. σκεδάννυμι). How the projection of images effects this we are not told, though we know it cannot be achieved by reproducing more originals of the same sort (31a, cf. *Republic* 597c). The plurality of the phenomena, then, simply seems to be an unanalyzed and possibly unanalyzable *given* in the Platonic metaphysics.

Cherniss, while wisely avoiding asserting that space is a principle of individuation for the phenomena, has suggested that the Demiurge of the *Timaeus* serves as a principle of individuation for the phenomena. This is an elegant variation on the recurrent view in the history of ideas that particulars can be individuated by their accidents and relations. Cherniss writes: "Since the spatial mirror is homogeneous and the Ideas themselves are non-spatial, the reflections in space would not be locally distinct; and the Demiurge is conceived as delimiting them by geometrical configurations, thus representing spatially the 'logical' distinctness of their non-spatial originals."[27] This interpretation runs up against the texts cited above which suggest a plurality of the phenomena in and of themselves prior to any demiurgic interventions, especially 52d–e, where it is claimed of the pre-cosmic receptacle that it receives shapes of earth and air, is qualified by all the other affections that go with these, has every sort of diverse appearance, and is filled with powers that are neither alike nor evenly balanced in any region of it. It is not possible, then, that there is in the pre-cosmos only one instance of each kind of image spread over the whole of space, so that the Demiurge can proceed to chop each up into many instances.

Rather I think it is safer to say that Plato does not have a principle of individuation for the phenomena and that probably he does not need one, since propositions all ultimately refer to Ideas, which are individuated by their unique conceptual content.

(3) One reason I think critics have tended to treat Plato's space as a principle of individuation is because they wish it to be what grounds the

[27] Cherniss (6) 255n18. For more discussion, see chapter 6, n50.

numerical unity of the subjects of basic propositions. They want the relation of visible property to space to correspond to and make possible the logical relation of subject–copula–(adjectival) predicate, and so to reproduce Aristotle's theory of predication. Zeyl is quite explicit about this.[28] He takes τοιοῦτον as describing "its referent as an *attribute* of something *else* . . . In other words, these terms ["fire," "air," etc.] are to be construed as logically (though not grammatically) *adjectival.* And this is precisely what the πιστὸς καὶ βέβαιος λόγος is: the construction of our nominal references to phenomena as adjectival descriptions of some basic, permanent subject worthy of that status. This subject is the receptacle, for only *it* can be designated as τοῦτο . . . Thus it appears that Plato's use of τοῦτο, τόδε, and τὸ τοιοῦτον is the direct ancestor of Aristotle's admittedly technical use of such locutions."[29] Problems for Zeyl's position here include his presuppositions (1) that Plato in this passage is talking about the way space determines the (logical) form of propositions and (2) that the form of "a stable and certain statement" is that of subject–copula–adjective. The latter presupposition is proven false simply by looking at the one and only sentence in the dialogue for which Plato claims absolute stability and certainty. The sentence reads: "So long as (the) two things are different, neither can ever come to be in the other in such a way that the two should become at once one and the same thing and two" (52c6–d1). The sentence is rather opaque, but at a minimum it is a proposition stating a relation between at least six terms (sameness, difference, oneness, twoness, simultaneity, and coming-to-be-in). The sentence cannot be reduced to a subject–copula–adjective form.[30]

In any case, that space is called a "this" is not a consequence of the term "this" picking out a numerically unitary substrate for phenomenal properties, since the criterion given for the designation "a this" is stability, not numerical identity (49d7–e4). The stability of the phenomena that allows them to be described by discourse that is *somewhat* stable (49b5) is a consequence not of their relation to the numerical identity of space or to a numerical unity which it is allegedly to provide them severally, but

---

[28] Zeyl (1) 146–48.

[29] Zeyl (1) 146–47, his emphasis.

[30] At a minimum, the sentence holds that Forms cannot be in the receptacle of space and the receptacle cannot be in the Forms, since a Form and the receptacle are each on its own fundamentally one in number. But the claim's neutral wording suggests a much wider range of applications. If, at the writing of the *Timaeus*, Plato were familiar with the contents of (what would become) the *Categories*, the sentence would be a critique of Aristotle's claim that an individual from a non-substance category can and must be in a substance. For Plato, this "being in" is impossible since, for the Aristotle of *Categories* 2, the substance and the individual non-substance are each by itself fundamentally one in number.

is explicitly the result of the relation of resemblance which they hold to the Ideas (29c). If the *Timaeus* suggests a logical form for propositions about phenomena—Lee has suggested the form "Redness, or whateverness, is here"—space does not enter into this logical form (if it does at all) as the source of sentences' stability; and so Plato's space does not, as does Aristotle's matter, serve as the objective correlate to the subjects of predication.

(4) As a result of not being a subject for predication, space for Plato is not, at least in Aristotle's sense, a substrate for change. If by change we mean a subject's having one predicate apply to it at time T_1 and another at time T_2, then since space is not a subject for predication, it is not that which abides through the change of predicates. It is true that the receptacle, as the gold analogy shows, does abide or remain the same while the phenomena change, but it does not, as the gold analogy (properly interpreted) also shows, enter into the process of change as a constituent of that which changes. We cannot point to the receptacle as a constituent of a phenomenal object and use it as the referent for determining that a phenomenal object is the same object even though (some of) its properties change. And so too Plato's space, unlike Aristotle's matter, is not potentially the contrary of any particular phenomenal object. Space is not, then, at least in Aristotle's sense, a substrate for change. Space provides a field across which images may flicker, but it does not provide any continuity to the images as they change.

(5) Aristotle does not explicitly state that matter is a principle of existence for the phenomena, but this seems to be strongly suggested by *Metaphysics* VII (1032a20–22, 1039b20–31) and by *De Caelo* 278b1–2, and it is implied by the doctrine that being and unity are synonymous in each of the categories *(Metaphysics* 1003b22–37), such that, since to be one in the substance category is to be one in number and so to be individuated by matter, matter is also the principle of being for the substance category, namely, it is that by which substances exist *simpliciter.* One might say that "to be" is "to be enmattered" for Aristotle. Plato's view might be analogous. Space is a principle of existence for non-substantial images. It is claimed at 52c that since a phenomenon is a "semblance of something else, it is proper that it should come to be in something else [namely, space] clinging in some way to existence on pain of being nothing at all" (Cornford). It is only as a principle of existence that Plato's space and Aristotle's matter seem to have a similar function.

Even the *descriptive* properties of space and matter are assigned for different reasons by each philosopher. That space for Plato is eternal is the consequence of its being a field or medium for images rather than of its being that which abides changes of predicates.[31] And Plato's space is

---

[31] 50b6–c1; 52a8–b1, note δέ, b1; and γένεσιν, b1, means "coming into existence" not "going through change."

characterless so that it offers no interference to the reception of images (50b–c, d–e, 51a7), while Aristotle's matter is characterless because it is that of which everything else is predicated while it itself is not predicated of anything (*Metaphysics* 1028b36–37, cf. 1029a22–23).

As a medium or field which is necessitated by the interpretation of phenomena as images, Plato's space serves neither as that out of which phenomenal objects are made, nor as a principle of individuation, nor as a subject for predication, nor as a substrate for change. We may safely conclude with Cherniss then that "in Plato's theory, there is no material principle even for sensible objects."[32]

[32] Cherniss (2) 23.

FOUR

The Gold Analogy in the *Timaeus**

Since the appearance of Cherniss' "A Much Misread Passage of the *Timaeus*," a considerable body of literature has arisen dealing with the translation and interpretation of *Timaeus* 50a4–b5.[1] Nevertheless, I do not think these lines have been correctly translated or interpreted. I translate the passage, with labels to facilitate references and to indicate the passage's structure:

> But we must try again to speak more clearly concerning this. For,
> IA: if a man, while forming out of gold every type of shape, never stopped remolding each into all of the rest and if someone indicated one of the shapes and asked "what is it?"
> IB: then, by far the safest answer with regard to truth would be "it is gold,"
> IIA1: but, the triangle and all the other shapes which come to be in <the gold>,
> 2: these should never be said to be (real),
> 3: since, indeed, they undergo change even while this is being asserted,
> IIB: but if, then, he is willing also to accept with some certainty <the answer, "it is> something of a certain kind," we should be glad.

I will offer first a brief overall interpretation of the passage and then will turn to a detailed analysis of each clause.

As with Cornford, Cherniss, Mills, and Cherry and in opposition to Lee,[2] I take the passage as a recapitulation and clarification of 49b–e; but, in addition, as Lee has noted, the passage must also be interpreted so that it serves as a transition to 50b5 ff. (ὁ αὐτὸς δὴ λόγος).[3] I take the main point of the much debated passage 49b–e to be the establishment of a contrast between the phenomena taken, on the one hand, as being in flux, about which, as the result of their being in flux, nothing whatsoever may be said, and the phenomena taken, on the other hand, as images. As images of Ideas, the phenomena may be said to be phenomena of a cer-

---

* Reprinted from *Phronesis* 23 (1978) by permission of Van Gorcum Ltd.

[1] Cherniss (9).

[2] Cornford (2) 180–85, Cherniss (9) 358–60, Mills 156–58, Cherry 8, Lee (3) 231.

[3] Lee (3) 231.

tain sort (τοιοῦτον).[4] The development of this double aspect of the phenomena (taken as in flux and taken as images of Ideas) is the main point, as I see it, of 47e–52c. I will argue that it is the main point of 50a4–b5.

Our passage develops this double aspect in two stages. First, it contrasts the phenomena as in flux (IA) with the receptacle (IB). Second, it contrasts the phenomena in flux (IIA3) with the phenomena as images (IIA1). Depending on which contrast is being considered we get a different answer to the "what is it?" question (50b1). On the one hand, insofar as the phenomena are in flux (IA), nothing whatsoever can be said of them. As in flux, the phenomena cannot be identified as to kind. Therefore, in pointing to a phenomenal object, all we may say in answer to the "what is it?" question is "it is the receptacle" or in the terms of the analogy "it is gold" (ὅτι χρυσός) (IB). On the other hand, insofar as the phenomenal object is a reflection, we may answer the "what is it?" question with "it is something of a determinate sort" (<ὅτι> τὸ τοιοῦτον, 50b4) (IIB). I will now offer a commentary to the various clauses of our passage to defend such a reading.

> But we must try again to speak still more clearly concerning this (50a4–5).

I take αὐτοῦ πέρι (a5) to refer back to the whole argument of 49b7–50a4 and not just, as Lee takes it, to 49e7ff.[5] Lee's reason for this is his claim that to τὸ τοιοῦτον of 50b4 is not a parallel usage to the earlier occurrences of τὸ τοιοῦτον at 49d5, 6, e5, 7. His sole reason for not seeing it as a parallel usage is that it lacks one of the temporal qualifiers (ἀεί, ἑκάστοτε, διὰ παντός) which on his (and Cherniss') reading of 49c–e accompanies each instance of τοιοῦτον.[6] However, I do not think ἑκάστοτε (49d5) and ἀεί (d7) can be taken with τὸ τοιοῦτον; rather, they qualify προσαγορεύειν (d6). If they were taken as single units with τὸ τοιοῦτον some participle like ὄν would be required to close the phrase in order to give the required sense: "that which is always such and such" (cf. ἀεὶ περιφερόμενον, 49e5).[7] In this case, though, τὸ τοιοῦτον by itself can and does stand as a shortened version of the fuller expression τὸ τοιοῦτον . . . συμπάντων (49e5–6). So Lee's reason for not seeing αὐτοῦ πέρι as referring back to the whole of 49b7–50a4 fails.

---

[4] My reading of the gold analogy, however, does not depend on this reading of 49b–e. As a clarification of the earlier passage the gold analogy must be able to stand on its own.

[5] Lee (3) 231.

[6] Lee (2) 5, Cherniss (9) 348n3.

[7] Even so, as Cherniss is aware, taking ἀεί and ἑκάστοτε with προσαγορεύειν is not itself disastrous for his overall interpretation of 49c–e. Cherniss (9) 348n2.

IA: For, if a man, while forming out of gold every type of shape, never stopped remolding each into all of the rest and if someone indicated one of the shapes and asked "what is it?"

IB: then, by far the safest answer with regard to truth would be "it is gold" (50a5–b2).

The topic of the protasis (IA) parallels the earlier εἰς ἄλληλα τὴν γένεσιν of 49c7 and alludes to the whole cycle of transformations of the primary bodies (earth, air, fire, water) described at 49b7 ff. Therefore, the concern of 50a5–b2, as of 49c7, is how phenomenal *types*,[8] not *individuals*, transform into one another. The πάντα σχήματα (50a5–6), then, denote not the extensions of all types of shapes, but just the types themselves. The passage is not concerned with the problem of how phenomena are individuated.[9] Though what is pointed out (δεικνύντος, 50a7) is necessarily an individual, the αὐτῶν ἓν of 50a7 refers not to the designated individual as an individual (i.e., as one among many individuals whether of the same shape or not) but to the kind of shape which the individual has in opposition to all other types of shape.[10,11]

Now the emphasis of the protasis (IA) rests squarely on the shapes taken as being in flux, since ceaseless transformation is the topic of the main clause, μηδὲν μεταπλάττων παύοιτο. The rest of the protasis consists of participial constructions.[12] Presupposing the phenomena to be in flux, the protasis goes on to reintroduce the problem of identifying phenomena according to kinds. This problem was first raised at 48b6–7. The problem is difficult to answer, we discover, because the phenomena are in flux (49b1–c7). The question τί ποτ᾽ ἐστί (50b1) alludes back to ὅτι ποτέ ἐστιν ἕκαστον αὐτῶν (48b6–7) and is equivalent to the interrogative τούτων . . . ἕκαστον ὁποῖον ὄντως (49b2–3).

Given that the phenomena are in flux, we cannot identify the type of individual at which we point. We can only say that the individual is the gold or the receptacle (IB).

---

8 Cf. ὁποῖον, 49b3, 4; ποῖον, 49d1.

9 As Lee would lead us to believe. Lee (3) 231.

10 Lee has correctly pointed out that the emphatic ἓν of 50a7 cannot be taken, as it is by Cherniss as the antecedent of ταῦτα (50b3). Lee (3) 221–22, Cherniss, (9) 358–59.

11 The ἕκαστα used in 50a7 to refer to phenomenal objects undermines Cherniss' reading of ἑκάστων (49d1) and ἕκαστα (49e4), which he takes as denying the possibility of such a usage. Cherniss (9) 352.

12 However, I do not accept Cherry's view that because the questioner is introduced in a genitive absolute he is introduced "on the side," and is therefore not obviously the subject of ἐθέλῃ (50b5). Cherry 8. Cherry suggests that if Plato meant this, he could have rewritten the sentence with the molder rather than the

But what sort of answer is this? It is not an adequate answer to the question τούτων . . . ἕκαστον ὁποῖον ὄντως (49b2–3), of which the question of IA is an abbreviation. Rather it is a proper answer for such a question as "what is the designated individual made of?" or "in what does it appear?" Now Plato could have just said, "Well, there is no answer to the 'what is it?' question." But he does not. I do not, therefore, agree with Lee, that Plato's saying the answer which he does give is μακρῷ πρὸς ἀλήθειαν ἀσφαλέστατον is meant to hold unconditionally and is to be contrasted with a less satisfactory answer given later μετ' ἀσφαλείας . . . τινος (50b4–5).[13] Rather, I take the answer "it is gold" (IB) to be the best possible answer only given the conditions laid out in the protasis (IA), namely, that the phenomena are being considered as in flux. To read the answer (IB) as Lee does is to accuse Plato of intentionally misleading the reader into supposing that the "what is it?" question has an answer which would make possible the articulation of the various phenomena and thus fulfill a promise made at 48b6–7 and reiterated at 49b2–3. I take this promise as being fulfilled in a general way in 49c–e, which I take to say the phenomena can be severally distinguished only as being images. I will argue that this view is expressed in the remainder (section II) of the present passage.

IIA1: But the triangle and all the other shapes which come to be in <the gold>,
2: these should never be said to be (real),
3: since, indeed, they undergo change even while this is being asserted,
IIB: but if, then, he is willing also to accept with some certainty <the answer, "it is> something of a certain kind," we should be glad (50b2–5).

I take the δὲ of 50b2 as contrasting the first answer to the "what is it?" question (IB) to everything after it (the whole of section II), not just as contrasting IB and IIA. The δὲ divides our passage in half. Section II, in turn, is divided into two contrasting sections (IIA and IIB) by the ἀλλ' of 50b4. As the result of (ἄρα, b4)[14] this second contrast, another answer (καὶ, b4) is given to the "what is it?" question. I take τὸ τοιοῦτον (b4) as grammatically equivalent to χρυσός (b2).

---

questioner cast in the genitive absolute. But Plato could not have done this without changing the emphasis of IA away from the *flux* of the phenomena, which he wishes to hold center stage.

[13] Lee (3) 223–28,

[14] Lee points out that ἄρα in a conditional protasis may mean "if, after all." But it need not mean this. See Denniston's remark: "Obviously different are passages in which εἰ and ἄρα are not connected in thought." Denniston 38.

In section II, though the phenomena are still treated as being in flux (IIA3), they are, in addition, treated as images. The expression ὅσα ἐνεγίγνετο recapitulates the expression ἐγγιγνόμενα . . . ἕκαστα αὐτῶν (49e7–8), which in turn looks forward to the designation of the various appearances in the receptacle as images of Ideas at 50c4–5 (τὰ εἰσιόντα καὶ ἐξιόντα τῶν ὄντων ἀεὶ μιμήματα).[15]

As images, the various phenomena, even though they are in flux, may be said to be of a certain kind. When we designate one of the shifting phenomena, we may say "it is something of a certain sort," insofar as it is an image of a stable original (IIB).

The gold analogy, then, has the following structure: IIA stands to IIB as IA stands to IB. Given the different conditions laid out in the A sections, two different answers are given to the "what is it?" question in the B sections. Insofar as the phenomena are in flux, the answer is "it is the receptacle," but insofar as the phenomena are images of Ideas, the answer is "it is something of a certain sort."

The gold analogy, then, is not to be construed along the lines of Cornford's interpretation of the contents of the receptacle. Cornford interprets the τοιοῦτον, which on both his and my reading of the gold analogy may be predicated of the phenomena, to mean "quality-like" and takes the gold analogy (and 49c–e) as contrasting the phenomena as being quality-like from 1) substances and 2) the geometrized primary corpuscles entertained at 53c ff.[16] Rather, in the gold analogy it is as images, not as 'qualities', that the phenomena may be said to be of a certain kind. The images are "such as" (οἷον) their originals. The analogy puts no restrictions on what kinds of images and originals there are.

In part we are glad we can make the second answer to the "what is it?" question because the very existence of the phenomena has been drawn into doubt (IIA2) as the result of their being in flux (IIA3). Cherniss and Lee have denied that the ὄντα of 50b3 can be taken existentially.[17] Their reasons are 1) that, so taken, ταῦτα ὡς ὄντα does not give an answer to the "what is it?" question and 2) that questions of existence are not what are under consideration in the analogy. However, in answer to 1), I do not, indeed, suppose that ταῦτα ὡς ὄντα is an answer to the "what is it?" question. Rather, <ὅτι> τὸ τοιοῦτον alone constitutes a second answer to the question. And against 2), I think a probing of the ontological status of

---

[15] I am in agreement with Cherniss on the textual parallels to ὅσα ἐνεγίγνετο. Cherniss (9) 359. However, Lee has shown that Cherniss' interpretation of IIA makes impossible demands on the text. Lee (3) 219–21. Cherry has shown that Cherniss' reading of IIA2 is impossible. Cherry 7, commentary note 7. See also n10 above.

[16] Cornford (2) 181–83.

[17] Cherniss (9) 360, Lee (3) 225.

phenomena fits into the context in three ways as part of the transition to the rest of the discussion of the receptacle. First, by reminding us that the phenomena are not real because they are in flux, the passage recalls the distinction with which the whole of Timaeus' discourse begins, namely, the distinction between τὸ ὂν ἀεί and τὸ γιγνόμενον μὲν [ἀεί], ὂν δὲ οὐδέποτε (27d6–28a1). As a result, we would expect that of which the phenomena are images to be equivalent to τὸ ὂν ἀεί, and indeed this equivalence is made at 50c5 (τῶν ὄντων ἀεὶ). If ὄντα (50b3) is not taken existentially, the assertion of 50c5 comes as somewhat of a surprise. Second, by denying existence to the phenomena as being in flux and by contrasting the phenomena as in flux with the phenomena as images, the gold analogy leads the way to the assertion that the shifting phenomena *as images* in the receptacle can "cling in some way to existence on pain of being nothing at all" (52c3–5, Cornford). Third, by raising the question of the ontological status of the phenomena in flux and by contrasting the shifting phenomena with that in which the phenomena are present, the gold analogy raises, in addition, the question of the status of the receptacle. This further question is taken up and answered in the immediately ensuing section (50b6 ff.). So we are justified in taking the ὄντα of 50b3 existentially.

Lee takes the ταῦτα of 50b3, as I do, to pick up τὸ τρίγωνον ὅσα . . . ἐνεγίγνετο.[18] However, Lee takes ὄντα predicatively, though as having no explicitly stated predicate. He interprets ταῦτα ὡς ὄντα as meaning "*they* are what the-object-that-the-questioner-points-to is."[19] This does not seem possible to me. Section IIA3, which Lee translates only parenthetically and on which Lee offers no comment, undermines Lee's reading of IIA2.[20] For, what is being pointed at by the questioner is one of the various types of shapes in flux (50a6–7). What Lee's reading of IIA2, then, amounts to is saying that the shapes in flux (IIA1&3) are not the shapes (σχήματα, 50a6) which are in flux (μηδὲν μεταπλάττων παύοιτο ἕκαστα εἰς ἅπαντα, 50a6–7). This is, of course, nonsense. Now if Lee wishes to get out of this problem, he could say that what is being pointed at is only the receptacle. Then, 50b3 (IIA2) would mean that the receptacle is not the various shapes in it. Lee seems to make this move.[21] Such a

---

[18] And so too does Cornford in his translation. In a note, however, Cornford offers an alternative reading which takes ταῦτα as the secondary object of λέγειν; Cornford suggests that "the contrast with τοιοῦτον following perhaps favors (this reading)." Cornford (2) 182n2. This *ad sensum* reading, however, misconstrues the gold analogy on the line which I have criticised above in discussing Cornford's interpretation of the analogy.

[19] Lee (3) 225, Lee's emphasis.

[20] Lee (3) 224.

[21] Lee (3) 223.

move, though, requires either that a second question has tacitly been introduced at IIA1, namely, "what, in turn, is the gold, or the receptacle?"[22] or that the first question is asking for the tautologous answer "the receptacle is the receptacle." Now Lee wishes to deny that two different questions are being asked in the gold analogy,[23] and the latter alternative is pointless. Lee's interpretation could succeed only if the "it" in the "what is it?" question were ambiguous in such a way that the two answers to the question could be 1) "it is gold" and 2) "it is not the shifting shapes." We need not, however, hoist such an ambiguity on Plato on my reading.[24]

I agree with Cherniss and Lee, against Cherry, that the questioner of IA is the subject of the "if" clause of IIB.[25] Mills has offered the novel interpretation that the gold itself is the subject which is said to receive what is denoted by τὸ τοιοῦτον. This reading jibes nicely with the description of the receptacle as all receiving (50b8, 51a7), but I do not think it is correct. First, it leaves unclear what, if any, relation exists between IIA and IIB. On Mills' reading IIB seems to be a *non sequitur*. Second, such a reading commits Mills to viewing the Ideas, which he takes to be the referent of τὸ τοιοῦτον, as being present in the receptacle.[26] However, as Cherniss notes, the Ideas are said emphatically and unequivocally not to enter into anything whatsoever (52a2–3, c5–d1).[27] Third, Mills can not give a satisfactory interpretation to μετ' ἀσφαλείας . . . τινος (50b4–5). Mills takes the phrase to mean "in such a way that we can be reasonably sure just which of the various characters (triangularity, etc.) the gold is receiving at any particular time."[28] Mills seems here to suggest that the

---

[22] To which Lee gives the answer, "it is not the various shifting shapes."

[23] Lee (3) 219–21. This denial is the basis of Lee's strongest argument against Cherniss' reading of the analogy.

[24] Aside from this problem with Lee's overall reading of the passage, Lee's admitted failure to find a role for the passage in its context, seems disastrous for his reading, in light of other readings which make the analogy intelligible in its context, which Lee himself has argued is a well-integrated whole. Lee (3) 230–31, (1) 343–49.

[25] Cherniss (9) 359–60, Lee (3) 223, Cherry (see n12 above).

[26] Mills 154, 159, 167.

[27] Mills is not unaware of this problem. Mills 159. He attempts to get out of the problem by saying that only the Ideas as images are present in the receptacle. He is correct in saying that only images are in the receptacle (50c4–5, 51a2, 52a4–7, 52c). However, the very reason Plato says the Ideas do not enter into anything at 52a2–3 is to differentiate the Ideas from images, after having first differentiated the Ideas from the receptacle by saying the Ideas receive nothing into themselves (52a2). It is not possible, then, for Mills to claim the images are aspects of the Ideas which make possible the presence of the Ideas in the receptacle.

[28] Mills 156n20.

receptacle offers a distorting resistance to that which it receives such that we are uncertain what we see when looking into the receptacle. It is clear, however, that the reason that the phenomena are hard to identify is not their mere presence in the receptacle but rather their being in flux (49b2–d3, 50a5–b1). Moreover, the receptacle is completely characterless exactly so it will exert no distorting resistance to that which it receives (50e). It is for this reason that gold is picked for the analogy with the receptacle. Gold is malleable, so it can receive and hold all shapes, unlike, say, a liquid, yet it offers no resistance to the various shapes which it receives, so it can constantly be shaped and reshaped, unlike, say, stone.[29] I think, therefore, we are justified in rejecting Mills' reading of IIB.

I suggest that the interpretation I have given for the gold analogy is the correct one. When we point to a particular phenomenal object, insofar as it is in flux we may say nothing about it. Given the condition that the phenomena are in flux, all we can say is that what we are indicating by our pointing is the receptacle. We recognize, though, that such a statement is not a suitable answer to a question which inquires into the τί or ὁποῖον of phenomenal objects. Further, as being in flux, a phenomenal object is on the verge of non-existence, so we are fortunate that the phenomenal object is also an image such that in pointing to it we can say that it is something of a certain sort and so give a proper answer to the "what is it?" question. That the phenomena are reflections of Ideas saves them from non-existence and complete unintelligibility.

Most critics have supposed that if Plato in the gold analogy and in the preceding lines (49b–e) permits τὸ τοιοῦτον to be predicated of the phenomena, then he is contradicting other of his writings, in particular 1) the *Cratylus* where at 439d8–12 it is said of what is always in flux that you can not say either that it is ἐκεῖνο or τοιοῦτον and 2) the *Theaetetus* where at 182c–183c (cf. 152d, 157b) all possible predications are denied of that which is in flux. Various solutions have been offered to get round this problem. Cherniss has suggested that the various occurrences of τὸ τοιοῦτον at *Timaeus* 49d–50b do not denote the phenomena, but some ontological stratum different from the receptacle, the phenomena, and the Ideas.[30] The introduction of a fourth ontological stratum, however, runs

[29] So correctly Johansen, who also correctly diagnoses the transience of phenomena, not the difficulty of individuating them, as the problem that the gold analogy is addressing. Johansen 122n8. But Johansen then slips back into viewing the gold in the gold analogy as a precursor to Aristotle's matter, that is, as a substrate for change. Johansen 134. Missing the point that it is the malleability of gold that is the relevant feature of the analogy, Zeyl takes Plato's choice of gold for the analogy as the chief evidence that the receptacle is being viewed as a material substrate out of which the phenomena are made. Zeyl (2) lxii.

[30] Cherniss (9) 361–63. Cf. Lee (1) 367. This position is defended by Silverman, (1), (2) 246–84.

up against Plato's emphatic and repeated assertion that he is dealing with only three strata.[31] The only item which the discussion of Necessity adds to the inventory of 27d–28a is the receptacle (49a1–6, 52a8).

Gulley and Cherry take τοιοῦτον as predicable of the phenomena in the *Timaeus* and take Plato as having abandoned a universal flux doctrine by the time of the writing of the *Timaeus.* The problem with this view is that both *Timaeus* 49b–e and the gold analogy start by presupposing that the phenomena are in a universal flux. The phrase ἀεὶ ὃ καθορῶμεν ἄλλοτε ἄλλῃ γιγνόμενον (49d4–5) summarizes the description given at 49b6–c7 of the flux of the phenomena, and the flux of the phenomena is reintroduced in the gold analogy both at 50a6–7 (in IA) and 50b3–4 (IIA3). In both 49b–e and the gold analogy, it is the flux of the phenomena which raises the difficulty of identifying the phenomena. If Plato did not believe that the phenomena were in flux, he would simply be shadow boxing in these passages; the source of the difficulty for answering the "what is it?" question would vanish under Gulley's and Cherry's analyses.[32]

On my reading of the gold analogy, we need not add to Plato's ontological commitments nor do we find a disparity in his thinking about the flux of the phenomena. The *Cratylus* and *Theaetetus* passages cited above hold that an object, whatever it may be,[33] insofar as it is in flux can have nothing whatsoever predicated of it. The *Timaeus* on my reading reaffirms this view. The phenomena insofar as they are in flux are not subject to any predications. However, nothing is said in the *Cratylus* passage about the phenomena's relation to the Ideas. Though the phenomena are asserted to be in flux (439d3–4), the status of the Ideas and so also the relation of the Ideas to the phenomena are left completely open questions possibly to be taken up at some later date (440b–e). Similarly, in the *Theaetetus,* whether or not one reads the dialogue as a *reductio* argument for the existence of Ideas, nothing is said of the relation of the Ideas to the phenomena. In particular, neither passage of the *Cratylus* and *Theaetetus* says anything of the phenomena being images of the Ideas. So the possibility is left open in these dialogues of distinguishing the aspect of the phenomena as in flux from the aspect of the phenomena as images of Ideas. The *Cratylus* and *Theaetetus* passages are compatible with the view that as images of Ideas the phenomena may be distinguished according to kinds and so be subjects of the predicate τοιοῦτον. That Plato in the *Ti-*

[31] Cf. Mills 154, 170.

[32] It is for this reason that I cannot accept Cherry's explaining away all the references in the *Timaeus* (and *Philebus)* to the flux of the phenomena by saying they are not to be read literally but rather "as emphatic ways of stating the inferior ontological status of γιγνόμενα." Cherry 5.

[33] In the *Cratylus* it is the Ideas themselves that are, for the sake of the argument, entertained hypothetically as being in flux.

maeus thought of the phenomena as having such a double aspect is clear from 50d1–2 and 52a4–7 where the images of the Ideas and that which is in flux are treated as being co-extensive.[34] The gold analogy, then, does not disrupt the unity of Plato's thought. Rather, in a few deft strokes the gold analogy clarifies and unifies Plato's thoughts on the possibility of making predications of the phenomena.

[34] These passages militate against Cherniss' view that the phenomena and the images of Ideas are not one and the same, but form two distinct ontological strata. Cherniss (9) 361–63. See also 50c4–5, 48e6–49a1 and 29b7–c3, where the phenomena in flux and images, likenesses, or imitations of Ideas are taken as extensionally equivalent.

FIVE

Remarks on the Stereometric Nature and Status of the Primary Bodies in the *Timaeus**

In the first section of this chapter, I argue for the view that the primary bodies (earth, air, fire, and water) of the *Timaeus* are present even in the pre-cosmic chaos (30a, 52d–53a) as stereometric particles prior to any intervention into chaos by the Demiurge. This formerly popular view has for the most part been dropped by Platonic scholarship since Cornford's broadside attack on it.[1] In the first section, I also argue that the nature of the Demiurge's role in fashioning the primary bodies has largely been misunderstood by critics. In the second section, I deal with the ontological status of the stereometric particles and of the elemental triangles out of which they are composed.

I

It is often supposed that if the Demiurge is the source of the order of the primary bodies (53b, 56c, 69b–c), then the primary bodies could not have existed in their geometric forms, as described at 53c ff., in the pre-cosmos. I wish to suggest, however, that the contents of the receptacle

---

* This chapter was first published in *The Platonic Cosmology*.

[1] Cornford (2) 159ff., especially 180–81, 182n4, 186n3, 190, 198–206. Notable holdouts for the formerly popular view are Crombie 219–20, 222–23 and Mortley 17 (see n16 below). And Zeyl appears to have accepted the argument of this chapter: "What are these 'traces' [of the four primary bodies in the receptacle]? Presumably they are collections of corpuscles consisting of various accidental combinations of irregularly shaped surfaces, combinations that fall far short of the artfully constructed polyhedra to be described later but coincidentally resembling them in some ways, and to that extent, behaving like them in tending to appear in different regions of the receptacle." Zeyl (2) lxvii. All of this is exactly right. The view is rejected by Broadie and Johansen, though neither seems wholly comfortable with the results of doing so. Contrast Broadie (2) 185 with Broadie (2) 188n17. Johansen leaves very unclear how pre-cosmic and post-creational transformations of the primary bodies are related. Johansen 124–27, especially 127.

even in the pre-cosmos are geometrical particles, the shapes of which, though, frequently deviate from the perfectly regular solids described at 53c ff. The role of the Demiurge is to bring deviate particles into accord with their paradigmatic Forms by eliminating any degree to which they fall away from their paradigmatic Forms, thus making them perfect instances of their corresponding Forms.

In the description of the actual construction of the stereometric particles (53c ff.), it is noteworthy that the Demiurge is not mentioned at all. Rather the constructions are cast in the first person plural. This seems to suggest not that the Demiurge constructs the particles, but that the particles are given in nature. That they are "constructed" at all is simply an aid for understanding their structure.[2] Note the description of the "constructed" particles as τὰ γεγονότα νῦν τῷ λογῷ (55d7). Perfect rather than deviant instances are described at 53c ff., understandably enough, for convenience of expression.[3]

The description of the Demiurge' s first interventions into chaos (53a7–c3) begins by suggesting that earth, water, air, and fire were all without measure (ἀμέτρως) in the pre-cosmos (53a8). Later it is said that the Demiurge produces measuredness (συμμετρία) in each of the primary particles (69b4, cf. μέτρον, 68b6). Plato's theory of the relation of measure to artistic production is laid out most fully in the *Statesman* (283c–287b). The measures of the *Statesman* are paradigmatic Forms.[4] It is by a craftsman's knowledge of measures (284c) that the products of the arts (284a5) are brought into being (285a1–4) and made good and fine (284b2). The Demiurge or craftsman looks to the measures or Forms, takes what is a degenerate instance of them, i.e., that which is subject to the more and the less, and makes, or at least attempts to make, a perfect instance out of it by the elimination of excess and deficiency. This is what the Demiurge is said to do at the start of the *Timaeus*, where the Demiurge is said to look to an eternal standard and bring the world into accord with it (28a, 29a), and so too, I suggest, this is the sort of activity performed by the Demiurge at 53b.

Critics tend to forget that what the Demiurge works on at 53b are images of Ideas.[5] The images may be degenerate instances of their Forms,

[2] The γένεσις of the particles mentioned at 53b8 refers not to their coming-into-being, but rather to their mutual transformations into each other, described in detail at 56c ff. (so also, γένεσις, 53e3, 54b7–8, d3).

[3] These considerations should allay Broadie's worry that irregular particles fail to be explicitly mentioned in the description of both the construction of the primary bodies and the contents of the pre-cosmic receptacle. Broadie (2) 190n36.

[4] Mohr (1). See also Cherniss (4) 130–31.

[5] The phenomena in flux (which the Demiurge takes over) and the phenomena as images are extensionally equivalent (29c1–3, 48e7–49a1, 50d1–2, 52a4–7).

and so may not be fully and properly called after their models (69b7–8), but are nevertheless recognizable as images. The images, though they are said to be in a state one would expect in the absence of divinity (53b3–4), are said to have some vestiges of their own natures (53b2) and indeed to participate somewhat in measure by mere chance (69b6), which I take to mean "by their mere presence in the receptacle as images of Ideas." I take these statements to mean that the primary particles existed as recognizable, though usually degenerate, forms of what they would become when the Demiurge eliminated their deficiencies and excesses—their departures from exact correspondence to their paradigmatic Ideas—and made them, each and all, completely regular.[6]

The Demiurge does not take over the phenomena in the pre-cosmos as some formless matter to which he then gives form, as though he were making a terracotta sphere from a shapeless glob of clay. Yet this is invariably how critics interpret 53b.[7] Such interpretations consciously or not read into the *Timaeus* Aristotle's form/matter distinction. But the Demiurge's action in eliminating deficiency and excess from degenerate particulars is more like straightening out a blade that has become warped than creating something new from that which is wholly formless. The Platonic maker is like a doctor who reduces an excessive temperature from a fevered patient or raises the deficient temperature of a patient with hypothermia. In both cases the doctor shifts the patient's temperature along a continuous scale of temperatures till it corresponds exactly to the temperature dictated by the Form or standard of health (*Phaedrus* 268a–b with 270b, *Philebus* 25e8, 26b6). At no point is the patient without a determinate temperature, though we would only be able to identify precisely what it is by reference to a standard, a thermometer. The doctor does not impose a determinate temperature where there was no temperature before. Rather he imposes the right temperature.

---

[6] Cornford's claim that the "vestiges" of fire, air (etc.) are qualities (e.g., heat, yellowness, brightness), which he takes to be constitutive of fire (etc.), rests on his understanding of 49b–50a, which he takes as establishing the contents of the receptacle as sensible qualities taken in contrast to substances in general and particles in particular. This also appears to be Johansen's view of fire and the like in the pre-cosmos. Cornford (2) 178–81, Johansen 124–27.

Further, just because Cornford successfully argues against Taylor that the receptacle does not operate like Democritus' sieve, he is not warranted to conclude that "with the Democritean sieve vanishes the last suggestion of discrete particles" as constituting the contents of the receptacle in the pre-cosmos. Cornford (2) 200–203, quote 202.

[7] For a paradigm case, see Vlastos (8) 27, 69–70 with 70n10. Vlastos writes: "He [the Demiurge] transforms *matter* from chaos to cosmos by impressing on it regular stereometric *form*." Vlastos (8) 70, my emphasis. Note Vlastos' description of that on which the Demiurge works as "formless material." Vlastos (8) 27.

Or consider examples of Demiurgic making from the *Timaeus* itself—the bringing into being of the rotation of the World-Body (34a) and the rotary and forward-proceeding motions of the heavenly gods (40b). The Demiurge does not "assign" to the World-Body the motion that is appropriate to it, namely, rotation, by taking a rigid body and giving it a spin, nor does he do so by taking some unmoving slush and stirring it around in a circle, thus introducing rotary motion where before there was no rotary motion nor indeed any sort of motion. No. He takes over a booming buzzing confusion of motions, all seven types of motion to which all bodies are subject in the pre-cosmos—motions forward, backward, up, down, right, left, and around—and then removes from the chaotic jumble the first six types of motion, which are called wandering motions. What remains after this excess is removed is the one standard, rational motion, to wit, rotation: "he took from [the World-Body] all the other six motions and gave it no part in their wanderings" (34a4–5, Cornford). Rotation was there all along, just buried in kinetic excess that needed to be removed. The heavenly gods have two motions—motion around and forward (40b). These gods are souls which reside in revolving spheres that are carried along a circular path: "But in respect of the other five motions he made each [god] motionless and still, in order that each might be as perfect as possible" (40b2–4). Goodness is achieved not by imposing characters onto a blank slate. Platonic goodness is achieved by the removal of excess and the bringing of a thing into accord with its standard.

Aristotelianizing critics of Plato's physics generally try to squeeze their interpretation of the Demiurge's actions out of the sentence ταῦτα πρῶτον διεσχηματίσατο εἴδεσί τε καὶ ἀριθμοῖς (53b4–5). Cornford translates: "The god [i.e., the Demiurge], then, began by giving [earth, water, air, and fire] a distinct configuration by means of shapes and numbers." This translation takes the datives as an instrumental usage, which admittedly has a *prima facie* plausibility given the presence of a verb of action and making. But I suggest that this translation is probably wrong and unfair, and is at least misleading. It is hardly fair to translate διεσχηματίσατο as "gives a distinct configuration," if at 50c3 the very same word, there used to describe the pre-cosmic contents of the receptacle, meant only to "form some kind of pattern, however vague in outline and irregular."[8] Rather the two occurrences of the term suggest that the contents of the receptacle already consist of configurations, which are then enhanced by the Demiurge's actions. It is hardly fair to translate εἴδεσι as "shapes," if at the start of the description of chaos μορφάς (52d6) had to be translated as "characters" because these two terms are supposed to

[8] Cornford (2) 185n1.

have the same meaning.[9] And yet if εἴδεσι is read instrumentally and translated neutrally and non-technically (in keeping with its many non-technical occurrences in the context) as "characters" or "characteristics," then the statement does not give the required Aristotelian sense, nor does it make much sense at all. For if the constituents of the pre-cosmos are characteristics or mere qualities, as they are on Cornford's account, it hardly makes sense to say in addition that the characteristics are shaped by means of characteristics.[10]

Moreover, taking the datives instrumentally does not even achieve the desired interpretative result, namely, that the Demiurge is the source of the geometric shapes of the particles. For it is not the case that doing *x* to *y* by means of *z* implies that *y* comes to have the properties which *z* has. If a sculptor works on a statue by means of a point and mallet, this does not involve the statue coming to have the properties which the point and mallet have. Therefore, even if the contents of the receptacle were worked upon "by means of shapes and number," they would not necessarily become shaped and numbered. Minimally, we would need some further details on the nature of the Demiurge's crafting of the primary bodies with the use of instruments in order for the Aristotelianizing account to succeed. If, for instance, a craftsman were described as fashioning something by molding or stamping it out of some material with the aid of a mold or stamp, the thing thus made would indeed have the shape (though sometimes in reverse) of the mold or stamp. But such saving details are never forthcoming here or elsewhere in the dialogue; εἴδεσι even together with ἀριθμοῖς can hardly be construed itself as describing or even implying the use of molds and stamps. And the Demiurge certainly is never described as operating like a pressman or one who fashions with the aid of molds, *even when* he is described as manipulating stuff-like materials in the first half of the dialogue (as at 35a–b, 41d). Indeed we never hear of the Demiurge using any tools in his crafting in the first half of the dialogue.[11] It would be strange for Plato suddenly to be populating the ontology of the second half with a new type of entity without clearly flagging that that was what he was doing, especially since the second half is written exactly to do that sort of thing—explain the introduction of a new ontological stratum, the receptacle (48e).

The datives of 53b5, I suggest, are datives of *respect*. This reading is strongly supported by 56c3–4 where the alterations wrought by the Demiurge on the primary particles are described as alterations *with re-*

---

[9] Cornford (2) 184, 188.

[10] Cornford (2) 180–81, 188,

[11] Indeed the only utensil of any sort that the Demiurge is said to use in the *Timaeus* is a mixing bowl (41d4), on which see "Extensions," section XI.

spect to number.[12] The Demiurge does not impose a number where there was no number before. In the particular case of 56b7–c5, he makes a number of particles which is too small into a number which is satisfactory for his purpose of making vision possible. In general, then, this means the Demiurge thoroughly shapes the images of earth, air (etc.), which fill the receptacle, with an eye to their (excessive or deficient) characters and numbers, which they possess prior to his intervention, and so he brings them into perfect accord with their Ideal models, whose existence is argued for at 51b–e.

Now this bringing into accord with a model occurs with the four primary bodies taken both severally and together (69b4–5). Such adjustment of the four primary bodies, taken together, with respect to number was described at 31b–32c in the discussion of the formation of the World-Body. Number for the primary bodies taken severally, though, involves the geometrical constituents of the primary particles (ἐξ ὅσων συμπεσόντων ἀριθμῶν, 54d4).[13] These minimally are the two types of elementary triangles "of which the bodies of fire and the rest have been wrought" (54b3–4).[14] It is these, then, from which the Demiurge eliminates deficiency and excess and which he brings into accord with their corresponding Ideas. The contents of the receptacle in the pre-cosmos, then, are primary particles which for the most part are themselves degenerate and consist for the most part of degenerate triangles.

This view is further supported if we look much later in the *Timaeus* where the organizing powers of the Demiurge and his ministers begin again to be absent from the primary corpuscles (see 89c1–4) as in the case of disease and death. With the onset of old age the primary particles, which are well-formed and consist of well-formed triangles (81b5–c1), begin to degenerate. They do not, however, degenerate by reverting to unformed qualities or formless matter. Rather, they simply become deviant, degenerate, and deficient triangles. They become weak (81c4) and

---

[12] Περί, though the woolliest of Greek prepositions, here (56c3), as the preceding lines (56b7–c3) show, must be taken to mean "in respect to."

[13] Cf. Cornford (2) 216n2.

[14] This undermines Taylor's undefended suggestion that the ἀριθμοί of 53b5 "are clearly the numerical formulae by which the different εἴδη are determined." Taylor (1) 358. Taylor takes εἴδη here in a technical sense to be "the geometrical shapes of the particles, which are about to be described: a sense of εἶδος which still survives in Euclid." Taylor (1) 358. However, the many earlier non-technical uses of εἶδος in the discussion of Necessity and in particular the use at 50e4 suggest rather that εἴδη at 53b5 means "characteristics" taken in the broadest of senses, by which it refers to the content which *any* image in the receptacle has as an image of a particular Form. The expression is not limited to denoting sensible qualities to the exclusion of shapes as Cornford would have it. Cornford (2) 188.

warped (82d6 with e2 and 73b6). Thus we must assume them also to be in this state in the pre-cosmos when the Demiurge is absent.[15,16]

II

What though is the status of the elementary triangles and the figures constructed of them? It is my view that the triangles are simply two dimensional plane figures which when forming regular solids enclose vacancy or empty space.[17]

There is a general hesitancy among critics to accept the view that plane figures are the ultimate constituents of the Platonic physical universe. There is a general feeling that geometrical figures in the phenomenal world must inhere in substances as an oval inheres in an egg or as a rectangle inheres in a desktop. This alleged inherence takes one of two forms. Either the triangles and particles are thought of as being made out of the receptacle[18] or the triangles are thought of as consisting of thin plates or lamina of matter, like flakes of mica.[19] But both such strategies betray a peculiarly Aristotelian prejudice.[20] The elementary triangles *as images* are on the same footing for Plato as any other phenomena. That is,

---

[15] Broadie fails to take into account this evidence of warped and weak triangles when claiming that the assertion that geometric but degenerate particles are present in the pre-cosmos is "gratuitous" and "makes no cosmological point." Broadie (2) 190n36.

[16] Another argument for this view is given by Mortley based on the theory of sense-perception in the *Timaeus*. Briefly, the argument is that the contents of the receptacle in the pre-cosmos cannot be visible properties which in the cosmos are *constitutive* of the primary particles, since in the cosmos sensible properties are epiphenomenal *consequences* of the shapes of the particles (61c ff.). One cannot get around this difficulty for Cornford's view by saying that the sensible qualities in chaos are not carried over in the cosmos, since they are, on Cornford's reading, constitutive of the stereometric particles. Therefore, the particles must exist, though in degenerate form, in the pre-cosmos.

[17] So Vlastos (8) 70, 90. However, since the triangles exist even in the pre-cosmos, we need not look on the Demiurge's actions as a "stupendous operation" on a par with a fairy turning "a pumpkin into a coach-and-four." Vlastos (8) 70n10. The Demiurge's actions would indeed be stupendous if he started with some three-dimensional "stuff" and transformed it into two-dimensional plane figures. But he does not. Rather he is like a carpenter who takes rough cut boards and makes them all a yard long by referring them to a yardstick, which he uses as a standard.

[18] Crombie 222 and Miller.

[19] For discussion, see Cornford (2) 229.

[20] See Taylor (1) 408–409.

they exist by appearing in and clinging to the receptacle (52c). There is no need, therefore, for the elementary triangles to inhere in substances or to be (made out) of some formless matter.

Cornford denies that the primary particles are "empty boxes—nothing but geometrical planes enclosing vacancy."[21] Cornford gives two reasons for this denial.

The first is that if the 'empty box' view were the case, 1) the particles would be "inanimate" and therefore, 2) there would be "no motion" in the phenomenal world.[22] However, the inference from 1) to 2) entails two hidden premises. First, the inference entails that the doctrine from the *Phaedrus* (245c–d) and the *Laws* X (896b) that soul is the cause and principle of all motion can be found in the *Timaeus.* But here, as in the discussion of the psychology of the *Timaeus,* Cornford by using the doctrine not as a possible suggestive *guide* for interpreting the dialogue, but as a *premise* in his interpretation, simply ends up begging the question of whether or not the doctrine is to be found in the *Timaeus.*[23] Second, the inference from 1) to 2) entails that soul is immanent in each and every thing which it moves. But this is not entailed in the principle of motion doctrine. It is perfectly possible to read the principle of motion doctrine as requiring only that soul proximately move some things, which in turn move other things which on their own have no direct contact with soul.[24]

Cornford's second reason for denying that the primary particles are "empty boxes" is that A) "space would be empty—a void partitioned by geometrical planes" and that B) "Plato's description throughout implies that the particles are i) filled with those changes or powers which are sensible qualities; and that they are ii) penetrated and animated by soul."[25] Against A), it should be noted that when Plato says that space is full, he means only that its contents are closely packed. This is all that is required at 58a and 80c where it is said that space is full.[26] When Plato says that space is full of powers (52e), this is like saying that a room filled with orange balloons is full of orange color. That space is full in the sense of being filled with closely packed particles which have unfilled interstices between them[27] and not in the sense that a tank is full of a liquid

---

[21] Cornford (2) 205.

[22] Cornford (2) 205.

[23] Cornford (2) 57.

[24] Such a view is strongly suggested by *Timaeus* 46d4–e2. See also chapter 6 below.

[25] Cornford (2) 205.

[26] So Vlastos (8) 90n34.

[27] That is, like a room filled with balloons such that, though there is quite a lot of space collectively in the interstices between the balloons, there is no one interstice in which an additional balloon could be placed.

such that there are no void interstices is not only sufficient to account for the descriptions at 52e, 58a, and 80c, but also is necessary for the explanation at 58b of how motion is maintained in the phenomena (58b4–5). The particles which fill space neither severally[28] nor collectively are a plenum in the sense of something which is continuous and infinitely divisible (at least) in thought.

Further, Cornford's independent reasons, i) and ii), for B) are false. It is not implied throughout the *Timaeus* that the particles are filled with powers or sensible qualities. First, this would clash with the epistemology of the dialogue.[29] Second, it is space as a whole that is said to be filled with powers (52e); nowhere is it *said* that the particles are filled with powers. Rather i) is an inference from Cornford's mistaken assumption that what are given stereometric form are powers and sensible qualities, such that one could say, for instance, of a brass sphere that it is filled with the brass out of which the sphere is made. Concerning ii), it is indeed the case that soul is extended through the whole of space and all of its contents (34b, 36e). But since soul is repeatedly given a non-material status in the *Timaeus,* it is hard to see how soul can be considered as filling up the "empty boxes." Certainly the stereometric shapes are not made out of psychic constituents. And yet "being made out of" seems to be Cornford's criterion for being called "filled with." Cornford's objections to the view that the primary particles are geometric planes enclosing vacancy, then, collapse.

That the primary particles are geometric plane figures enclosing vacant space is necessitated by the description of the transformation of the primary particles (56c–57c). Various particles of different kinds, by grating with each other,[30] cut (57a2) or crush (56e4–5) each other such that they are dissolved (56d2) into their constituent triangles (τὰ μέρη, 56d4), which drift about (φέροιτ', 56d2) until they form again with themselves (56d4–5) or other triangles into stereometric particles (56d5–e1). It is necessary for Cornford to claim that Plato's description of the transformation of the primary particles, which involves fragments of particles drifting about, "cannot be taken literally."[31] Cornford goes so far as to call the description an "absurdity."[32] Cornford, however, fails to give the description a non-literal reading and tries to mask this failure by saying: "If we go behind Plato's description and ask after the 'real' nature of the

[28] In this they differ from Democritean atoms.

[29] See n16 above.

[30] This is a result of their moving in place (56e4) and yet being closely packed. Plato's primary corpuscles do not collide having first moved through open space, as do Democritean atoms.

[31] Cornford (2) 229.

[32] Cornford (2) 274.

process of dissolution and recombination, it is doubtful whether we can expect any certain answer."[33] Rather Cornford tries to give an analysis of the transformation of the particles which explains away the possibility that "the surfaces of any solid body can drift about by themselves."[34] Cornford's suggestion, based on Plato's alleged appeal to irregular geometrical solids in the explanation of odors (66d), is that we are "to understand that the triangles, on the way from one regular form to another, at every moment compose a series of irregular solids of intermediate size."[35] This analysis fails, however, if we look at the second inter-specific transformation which Plato asserts as occurring, namely, the transformation between two particles of fire (two four-sided pyramids) and a single particle of air (an octahedron) (56e1–2 with 57b3).[36] Now, two fire-pyramids of four sides each may butt up against each other such that one side of each completely overlaps one side of the other. We then would have a double pyramid, an irregular solid with six exterior faces, the very sort of intermediary solid which Cornford has in mind.[37] However, this six faceted solid cannot be further transformed into an octahedron consisting of just the original eight faces provided by the two pyramids, unless *at a minimum* each of the pyramids is "unfolded" along three of their six "joints" thus,

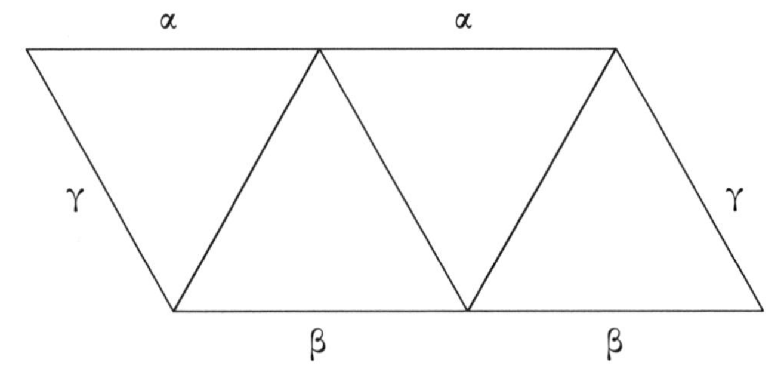

and the two unfolded figures are then (easily) recombined edges-to-edges into an octahedron.

Since then the various regular solids must come unjointed to account for the transformations of the primary corpuscles, the primary corpuscles must be considered as vacant regular solids made up of two-dimensional plane triangles.

---

33 Cornford (2) 229.

34 Cornford (2) 230.

35 Cornford (2) 274. Vlastos has convincingly argued against Cornford's account of the alleged presence of irregular solids in Plato's theory of odors. Vlastos (7) 366–73.

36 We may consider the surfaces of these figures as all of the same grade (57d4).

37 Cornford (2) 274.

SIX

The Mechanism of Flux in the *Timaeus**

I wish to argue that in the *Timaeus* Plato views the phenomena in and of themselves as a positive source of evil, moving erratically without psychic causes whether rational or irrational, direct or indirect; and so I wish to suggest that the *Timaeus* is inconsistent with the view expressed in the *Phaedrus* (245c9) and *Laws* X (896b1) that soul is ultimately the source of all motions, a corollary of which is that the phenomena *qua* phenomena are not a positive source of evil.

In 1939 Gregory Vlastos felt the need to apologize for adding to the enormous body of literature on the problem of the sources of disorderly motion in Plato's writings. Since that date interest in the topic has continued to generate a large body of literature. And yet, no unified consensus of opinion on the issue has emerged. Though the literature on the sources of evil problem is large, critics have tended to fall into five fairly well defined camps, four of which try to reconcile the principle of motion doctrine from the *Phaedrus* and *Laws* X with the existence, or at least apparent existence, of motions without psychic causes in the pre-cosmic period of the *Timaeus* myth (or in that which the pre-cosmic period represents on non-literal readings of the myth).[1] Three of these four camps claim that the existence of motions independent of psychic causes is indeed only apparent.

One of these three groups of critics claims that the World-Soul of the *Timaeus* directly causes erratic motions. But since the World-Soul also causes orderly motions, the whole soul cannot cause erratic motions. So it is claimed that part of the World-Soul, an irrational part, directly causes erratic motions, and that another part causes orderly motions. And thus, a psychic cause is found for all motions. This is Cornford's view, stated much in this deductive fashion more as an attempt to save apparently discrepant texts from contradiction than as an interpretation of any texts.[2] The *Timaeus* has been combed and no traces of irrationality in the World-Soul have been found. Cornford's view has not been an

* Reprinted from *Apeiron* 14 (1980) by permission of the editors.

[1] This chapter is neutral on the issue of whether the creation act in the *Timaeus* is to be read literally or analytically.

[2] Cornford (2) 57, 176–77, 205, 209–10.

especially popular one. It has been accepted only by Morrow, and has been extensively criticized by Vlastos, Skemp, Meldrum, Robinson, and Clegg.[3]

A second alternative which tries to find a psychic cause for erratic motions is to admit that the World-Soul is wholly rational, but to posit an irrational psychic force over and largely against the rational World-Soul as the source of erratic motions. This view has its roots in Plutarch[4] and was advocated in the twentieth century by Skemp,[5] Dodds,[6] and Clegg. A major objection to this view is that it is specifically entertained and rejected in the *Statesman* myth (270a1–2).[7]

The third view which finds a psychic cause for erratic motions in the *Timaeus* claims that the demiurgic aspects of the World-Soul are the cause of both orderly and disorderly motions, but of orderly motions directly and of disorderly motions indirectly as inadvertent but inevitable spin-off effects from rationally caused orderly motions. This is Cherniss' position,[8] which has been defended by Tarán.[9] This view has been attacked by Festugière, Herter, Vlastos, Skemp, Easterling, and Robinson.[10]

A fourth view, which is enjoying a rising popularity among critics, takes soul to be the cause of all motions *only* in the formed cosmos and not of the disorderly motions of chaos. This view simply dissolves the problem of there being any discrepancy between *Laws* X and the *Timaeus* by claiming these two texts are not talking about the same subject and therefore are not even in apparent contradiction. This view was ad-

---

[3] Morrow 437; Vlastos (2) 391–92; Skemp (1) 78–82; Meldrum 65–67; Robinson (1) 78–82; Clegg 53–54.

[4] Plutarch, *On the Psychogony in the Timaeus* in C. Herbert (ed.), *Moralia* 6.1 (Leipzig: Teubner, 1954) 1014a–1015f.

[5] Skemp (1) 74–78, 82–84, 112. Against Skemp, see Robinson (1) 96–97.

[6] Dodds 115–16.

[7] Skemp is not unaware of the problem this text presents and seems in dealing with the *Statesman* myth to push the irrational psychic force back within the World-Soul proper, but he leaves the whole issue of the status of his irrational World-Soul somewhat vague. Skemp (1) 26, (2) 107. Robinson suggests the obscurities surrounding the status of irrational psychic factors are endemic to the text of the *Statesman* myth and "point to Plato's discomfort about the whole matter." Robinson (1) 137. Clegg fails to take *Statesman* 270a1–2 into account. This omission is surprising and likely fatal for Clegg's account of the *Timaeus,* which is based almost entirely on an attempt to show that the *Timaeus* can be brought into doctrinal accord with an alleged coherence of Plato's thought on psychophysical matters in other dialogues ranging from the *Apology* to the *Laws.*

[8] Cherniss (1) 444–50, (6) (10) 458–61.

[9] Tarán 386–88.

[10] Festugière III, xii–xiv; Herter (1); Vlastos (3); Skemp (1) 148–51; Easterling 26–30; Robinson (1) 61–68.

vanced in 1939 by Vlastos, only to be abandoned by him in 1964, but has been accepted by Hackforth, Easterling, and Robinson.[11,12]

These four positions exhaust the interpretations which enable one to hold Plato as having a consistent view of physical causality between the *Timaeus* and the *Laws* X. There remains a fifth group of critics who hold that Plato is not consistent in his views on physical causation. These critics believe that the erratic motions of the pre-cosmos in the *Timaeus* do not have psychic causes and that according to the *Phaedrus* and *Laws* X they ought to. This view was advanced in the nineteenth century by Bäumker[13] and more recently by Festugière, Herter and Vlastos (1964).[14,15] It is with these critics that I have the most sympathies on this issue and whose views I will defend to the extent that I will try to show that the motions of the pre-cosmos are purely mechanical in origin.

To this end I argue, based on a reading of the description of chaos (52d–53a) (section I) and the description of the motion of the primary particles (58a–c) (section II), that the flux of the phenomena is the result of the geometrical nature of the particles.[16] I argue that Plato advances a theory of objective weight which is based on the geometrical construction of the particles. The different objective weights of different kinds of

---

[11] Vlastos, 1939, (2) 397; Vlastos, 1964, (3) 414–19, (2) 396n4; Hackforth (3) 21; Easterling 31–38; Robinson (1) 93–97.

[12] For criticism of this view see Cherniss (1) n362, Herter (1) 343, and chapter 8.

[13] Bäumker 145–48.

[14] Festugère II, 117–31; III, xii–xiv; Herter (1) 343–46; Vlastos (3) 414–19.

[15] Meldrum should perhaps also be ranked here, as he claims the model of soul from the *Phaedrus* and *Laws* X (which he calls the "Divine Animal" model) is not wholly compatible with the model of soul from the *Statesman* and *Timaeus* (which he calls the "Divine Workman" model). He, however, tends to treat the corporeal only as an inert, passive, or negative source of evil, that is, he treats the corporeal only as a stumbling-block to the organizing activities of the Demiurge, but not as itself an actively disruptive source of disorder.

[16] I accept Bäumker's thesis that the primary particles exist in the pre-cosmos in their stereometric forms. Bäumker 131–32. This view has largely been dropped by Platonic scholarship since Cornford's broadside attack on it. Cornford (2) 159 ff., especially 180–81, 182n4, 186n3, 190, 198–206. For some decisive arguments against Cornford's denial of this view see: Crombie 219–20, 222–23; Schulz 87–113; and Mortley. See also chapter 5 above.

The forward looking reference at 48b7–8 (which is governed temporally by πρὸ τῆς οὐρανοῦ γενέσεως of b3–4) to the elemental triangles at 53d4 ought to be by itself sufficient to establish Bäumker's thesis. But in any case, for the purpose of this chapter it is sufficient to note that 52d–53a and 58a–c must describe contemporaneous events, since the conditions described at 58a–c are introduced to avert the stagnation that would result from the complete separation of phenomenal kinds entertained at 53a1–2, 4–7 (with 57e2–3, 57e6–58a1).

particles give the different primary particles a natural buoyancy, an inherent actualized tendency for the particles to separate according to kinds. However, as the result of the formal properties of space and the differing efficiency with which different geometrical particles can be packed in space, the natural buoyancy of the primary particles indirectly causes the particles to be thrust back together again, and so the flux of the phenomena is caused and maintained by purely mechanical means.

I. THE DESCRIPTION OF "CHAOS": 52d–53a

The passage 52d–53a has been taken by some critics as itself a proof-text for the view that the phenomena by themselves without psychic intervention are in a universal flux or state of chaos. Though I agree with this view, I do not think that 52d–53a by itself constitutes a proof-text for it. This view can only be gleaned by taking 52d–53a together with the description of motion and rest at 57d–58c. Indeed I wish to argue that far from suggesting a chaotic state, 52d–53a only describes an inherent actualized tendency of the four types of primary bodies to separate into different regions—to be buoyant. I will argue later that only when this tendency of the four primary bodies to separate is taken in conjunction with the ability of the kinds to transform into one another do we end up with a description of universal flux or chaos (at 58c).

The passage begins with a recapitulation of 48e ff. Being, space, and becoming are said to exist even before the ordered universe (52d3–4). The receptacle is said to be made watery and fiery, to receive the μορφάς of earth and air, and to be affected by whatever other affections (πάθη) go with these (52d4–el). This phrasing parallels 51b4–6. In the present passage, it is likely that μορφάς carries at least in part the sense "shapes." Since the πάθη which follow upon the μορφάς are sensible properties (52e1), the passage foreshadows the epiphenomenalism of 61c–62c, where sensible properties are a consequence of the shapes of the stereometric particles. Cornford, who takes the μορφάς themselves to be sensible properties, does not and can not give an adequate explanation of πάθη.[17] He takes the πάθη to be just other sensible qualities. It is not clear though on this reading that τούτοις (52d6) has an intelligible antecedent. On the one hand, we might expect that τούτοις refers to earth, air, etc., yet sensible qualities are not *consequences* of these on Cornford's reading, but rather constituents of these. On the other hand, τούτοις cannot refer to μορφάς, since Cornford takes πάθη and μορφάς to denote the same things.[18] So one would not follow upon the other.

---

[17] Cornford (2) 199.
[18] Cornford (2) 198n1.

The point of this recapitulation of earlier material at the start of the present passage, I suggest, is that what is described in 52d–53a is simply a consequence of the expanded metaphysics of 48e ff., in which neither the Demiurge nor psychic forces of any sort are involved. What is described is simply the result of the projection of the images of Ideas into the mirror of space.[19] This argues against those who find, whether on a literal or non-literal reading of the *Timaeus* myth as a whole, psychic forces for the motions described in 52d–53a.

The first new bit of information we receive in this section is that the receptacle is filled with powers (δυνάμεις) which are neither alike nor evenly balanced (52e2–3). What are these powers? Robinson interprets them to be "forces," which he takes to be actual, not just potential, motions, and which he assumes to exist independently of material objects and, in particular, material particles.[20] I take "powers" in a much broader sense (with Cornford) to be applicable both to optics and kinematics and to suggest both the ability to act and to be acted upon and the actuality of acting and being acted upon.[21] As such, the powers refer to the interactions of the four primary kinds both with each other (as in the description of the transformation of the primary particles) and with an observer who is implicitly hypothesized as being present at 52e1 (cf. 61c–d). That the term δυνάμεις seems to serve as a transition from the mention of perceptible properties of the phenomena (52e1) to the kinematic properties of the phenomena (e4) suggests this is the correct interpretation of the term. Further confirmation comes from 56c4 where δυνάμεις are specifically not restricted just to motions, but include every aspect of the phenomena including visible properties (56c2–3). Sensible powers or πάθη are consequences of the shapes of the primary particles. I will be arguing (section II) that kinematic powers are also. The presence of powers in the receptacle does not rule out, as Robinson supposes,[22] that the powers belong to particles. Indeed, insofar as our present passage is concerned with the *interactions* of the phenomena with the receptacle we would expect the aspect of the phenomena as powers, rather than that of which the powers are powers, to be the focus of attention.

Now these powers are said to be neither alike nor evenly balanced and as a result there is no equipoise in any region of space (52e2–4). Further, because the receptacle is thus weighted differently everywhere, it is shaken by the powers and in turn anew (αὖ πάλιν) shakes them (52e1–5).

[19] This is especially so, since the διά of 52e1 shows the participles of 52d5–6 to be causal.

[20] Robinson (1) 94.

[21] Cornford (2) 199, (1) 234–38.

[22] Robinson (1) 94.

Is the shaking of the receptacle to be taken literally or figuratively? If literally, what role does it play in the economy of the present passage and the discussion of Necessity as a whole?[23] If figuratively, of what is the shaking a symbol? Critics have offered a diverse, but by no means exhaustive, array of possible answers to these questions.

Three basic views have been taken by critics. First there are critics that 1) take the shaking of the receptacle literally and 2) take the shaking to be the proximate efficient cause of the flux of the phenomena.[24] Such critics include Crombie, Cherry, Robinson, and Zeyl.[25] Second, there are those critics represented by Cornford and Cherniss who take the shaking figuratively. These critics divide, however, on what the shaking stands for. On the one hand, Cornford identifies the shaking with what he takes to be its efficient cause, an irrational psychic force.[26] On the other hand, Cherniss identifies the shaking with "the continual flux of phenomena" itself.[27] I will now try to show that these various interpretations are mistaken.

On the issue of a literal versus non-literal reading of the shaking of the receptacle, I agree with Cornford and Cherniss, though not for the reasons which they give for a non-literal reading. Cornford's reason is that all motion for Plato requires a psychic initiator and since there is no psychic initiator for the shaking of the receptacle, the description of the shaking is not to be read literally but rather as a spur for the reader to search for some psychic initiator for erratic motions.[28] Aside from beg-

---

[23] Literal readings are definitely on the rise in the twenty-first century. Zeyl (2) xxxv, lxiii, lxxi, Broadie (2) 181–84, Johansen 130–30. Johansen comes asymptotically close to seeing that the shaking of the receptacle read literally has no function in the Platonic system, but balks.

[24] View 1) does not in itself entail view 2), since the shaking of the receptacle could be a maintainer, transmitter, or transformer of motions which exist independently of the shaking. This is the view of Johansen and Broadie. For Johansen, the shaking of the receptacle simply amplifies what the primary bodies in it are already doing on their own—sorting into discrete concentric bands in a determinate order (from the center of the universe to its rim: earth, water, air, fire). Johansen 126. For Johansen it appears that the shaking of the receptacle performs no distinctive task. For Broadie, the shaking of the receptacle amplifies *and concentrates* the motions that are in it prior to its shaking: The receptacle "picks up the faint signals of motion emitted by the minute flecks and spots that are everywhere in it, and returns them amplified and concentrated so as to mass the different signal-senders into definite empirical realities." Broadie (2) 184.

[25] Crombie, 219, 221; Cherry, 8; Robinson (1) 94; Zeyl (2) lxvii, lxxi.

[26] Cornford (2) 209–10, so too Skemp (1) 58–59.

[27] Cherniss (1) 444–45. This is a crucial premise for Cherniss' later interpretation of 57d–58c. See Cherniss (1) n393.

[28] Cornford (2) 209–10.

ging the question on the nature and scope of psychic causation, this interpretation misreads the analogy which compares the shaking of the receptacle to the shaking of a winnowing-basket (52e6–7). Cornford writes: "Accordingly he [Plato] has to help out an impossible situation by comparing the receptacle to a winnowing-basket which both shakes and is shaken, leaving in obscurity the fact that, as surely as the basket needs someone to shake it, the bodily needs some soul before any motion can occur."[29] Cornford simply misreads the relevant detail of the analogy. The relevant detail, as the lines immediately following the introduction of the analogy show (53a1–7),[30] is that the contents of the receptacle, like the contents of a winnowing-basket, sort out according to kinds. The shaker is an irrelevant detail of the metaphor and is not only not mentioned in the description but is ruled out as the cause of the shaking, since the cause of the shaking is said to be the result of the unequal and unbalanced powers, which in turn are the images of Ideas present in the receptacle (52d4–e3).

Moreover, Cornford overlooks the expression αὖ πάλιν of 52e4–5, which entails that the contents of the receptacle are moving quite independently of the shaking of the receptacle. The contents move and *only then* in reaction to their motion does the winnowing-basket move—that's the literal story Plato spins here (52e4-5).[31] The expression entails that τὸ μήθ' ὁμοίων . . . ἰσορροπεῖν (52e2–3) describes an active rather than a static state.[32] In this case the shaking of the receptacle is not the proximate efficient cause of physical motion. The absence of a shaking agent in the description of the receptacle as a winnowing-basket at 52e is a significant difference from the reintroduction of the metaphor of the receptacle as winnowing-basket at 88d–e. However, against Cornford and those who see the shaking as the proximate cause of the flux of the phenomena, the *result* of the shaking both at 52d–53a and 88d–e is not chaotic motion but the orderly separation of phenomena according to kinds.[33]

---

[29] Cornford (2) 209.

[30] The καί of 52e7 is likely explicative.

[31] In his commentary, Zeyl disguises, by reversing, the order of motions described here. He says: "Prior to [the Demiurge's] intervention there were only 'traces' of fire and the others . . . in 'the wetnurse of becoming', the 'Receptacle' that both moves them and is moved by them." Zeyl (2) xxxv.

[32] See note 41 below. This argument also holds against those literalists, like Zeyl, who view the shaking of the receptacle as the proximate efficient cause of physical motions. Zeyl (2) xxxv, lxvii, lxxi.

[33] It is surprising, especially given the description of the separation of kinds at 32b–c, that Skemp can view the separation at 52e–53a by itself as an "acosmic tendency." Skemp (1) 59.

Cherniss claims the shaking of the receptacle is not to be taken literally as the result of A) his identification of the shaking of the receptacle with the pre-cosmic chaos and B) his claim that the pre-cosmic chaos itself is "mythical."[34] I wish to continue here to suspend judgment on B), but will for the sake of the argument give Cherniss this premise. Now if A) is true, and given B), then indeed C) the shaking of the receptacle is not to be taken literally. However, if A), then also D) the separation of phenomena according to kinds is due to the flux of the phenomena. Cherniss actually claims this and it is crucial to his interpretation of 57d–58c (see section II).[35] This interpretation (D), however, collides with the analysis of the receptacle given at 57c and at 88d–e. At 57c the fact that the receptacle sifts the phenomena according to kinds is given as a *reason* (γάρ, 57c2) for the flux of the phenomena (described as all the phenomena constantly interchanging their regions, 58c1–2) rather than as a *result* of the flux of phenomena. Further, at 88d the phenomena are assumed to be in flux on their own independently of their interaction with the body taken as a receptacle (88d4)[36] and yet they do not sort out according to kinds. Indeed it is in order to effect this sifting that the shaking of the receptacle is required and so a shaker is introduced at 88d–e. Therefore, the flux of the phenomena is not being viewed as the cause of the sifting of the phenomena according to kinds (therefore not-D).[37] And so the shaking of the receptacle is not equivalent to the flux of the phenomena (by *modus tollens,* not-A), and so too then Cherniss' argument for a non-literal reading of the shaking of the receptacle fails.

Moreover, that 1) 57c takes the flux of the phenomena to be the result of the separation of kinds, 2) 88d–e portrays the phenomena as being in flux independently of the shaking of the receptacle, and 3) 88d–e requires the introduction of a shaker to initiate the sifting of the kinds, suggests that at 52d–53a the shaking of the receptacle is not being viewed respectively as 1) a transmitter, 2) a maintainer, or 3) a transformer of erratic motions.

Neither Cornford's nor Cherniss' arguments for a non-literal reading of the shaking of the receptacle succeed. A better argument for a non-literal reading can be based on the characterlessness of the receptacle. The receptacle is said to be completely characterless so that it will not interfere with the reception of impressions from the Ideas (50b–c, 50d–51a). That is, the receptacle has no properties which distort the images it receives. The receptacle, of course, has the formal or external properties

[34] Cherniss (1) 444.

[35] Cherniss (1) n393.

[36] Note τὰ περὶ τὸ σῶμα πλανώμενα παθήματα (88e2).

[37] Therefore, 53a cannot be taken as asserting this, as Cherniss claims. Cherniss (1) n393.

of being eternal (52a8) and being one (52d1), but these properties do not influence what images it receives or the properties of the diverse images. Now Plato seemed to have believed that there are Ideas of motion (especially *Republic* 529d, *Sophist* 254b–d), so any motion which the receptacle had would influence its ability to receive undistortedly images of these specific Ideas.[38] And it goes almost without saying that the shaking of the receptacle would distort images of the Idea of rest, which plays such an important part in the *Timaeus'* litter-mate, the *Sophist* (254d). But more generally, since the motions of the contents of the receptacle influence what types of images are present in *it* at any moment and in any place (49b–c, 56c–57b, 58a–c), if the receptacle has a motion of its own which it sets up in turn in its content, it would disrupt the reception of *all* the images in it. Since any such distortion or disruption of the reception of images is repeatedly denied of the receptacle (50b8–c2, 50d–51a, and probably 50a2–4), it is reasonable to hold that the shaking of the receptacle is not meant to be read literally.[39]

The same conclusion could be reached for other reasons which we have established. If the shaking of the receptacle is neither an efficient cause of motion nor a maintainer or transmitter of motion, then it would not seem that the receptacle is meant to move at all.

If the shaking of the receptacle is, then, to be read non-literally, what does it stand for? We have seen that it cannot be identified with its *cause,* whether the cause is soul or the flux of the phenomena itself.[40] Nor, as we have seen, can it be identified with the flux of the phenomena whether or not the flux is its cause. Rather, I suggest, if the shaking is to stand for anything, it is to be identified with its *effect,* namely, the sifting of the phenomena according to kinds. More precisely, the shaking of the receptacle and the comparison of this shaking to that of a winnowing-basket is a device by which Plato is able to analyze the flux of the phenomena into components or aspects.[41] The component which the shaking

---

[38] Given Plato's epiphenomenal theory of sense perception (45b–e, 61c ff.), it seems likely that Plato is treating sensible qualities themselves as though they were motions. See the discussion of δυνάμεις above; cf. *Theaetetus* 155d–157b and Cornford (1) *ad loc.*

[39] That the receptacle shakes and yet is supposed to be characterless is one of the reasons proposed by Lee for severing 48e–52d from the ensuing description of chaos. Lee (1) 349–50.

[40] This is so since the flux does not cause the sifting of phenomenal kinds.

[41] I take the expression τὸ μήθ' . . . ἰσορροπεῖν (52e2–3) as portraying the flux of the phenomena even before the shaking of the receptacle is mentioned. The reasons for this are 1) the αὖ πάλιν of 52e4–5 entails that the contents of the receptacle are already in motion, 2) the "saving" of the flux of the phenomena (58c1–4) is viewed as tantamount to saving the heterogeneity (μήθ' ὁμοίων) mentioned at 52e2, 3) the δυνάμεις of 52e2 are, as we have seen, meant to include mo-

analyzes out of the flux is the sifting of the phenomena according to kinds.[42] Given a non-literal reading of the shaking of the receptacle and given that the shaking does not stand for either the cause of the shaking or the flux of the phenomena, the drift of the phenomena according to kinds is being viewed as occurring spontaneously without a direct or indirect psychic cause, whether of a rational or irrational sort.[43]

What the shaking of a winnowing-basket does is to actualize the natural buoyancy of grain. In the case of a winnowing-basket buoyancy is impeded and kept potential by friction. In 52d–53a the buoyancy of the contents of the receptacle is actual, like air or oil passing through water.

---

tions, and 4) only when so read is 52d–53a analogous in relevant respects to 88d–e where the phenomena are in flux prior to the shaking of that receptacle. Now the causal participial phrase ἀνωμάλως πάντῃ ταλαντουμένην (52e3–4) might seem to suggest that the contents of the receptacle are static and that they cause the receptacle to shake just as weights placed irregularly on an otherwise equally balanced, centrally suspended platter would cause the platter to wobble and shake. However, I do not think the participial phrase is to be read literally. The reason is that neither of Plato's analyses of weight (56b1, inherent weight and 62a–63e, perceived weight [see section II]) entails the receptacle, nor does it make sense on either analysis that the receptacle could be weighted differently by the phenomena in it, any more than a mirror is differently weighted by the various images it receives. The participial expression is just a means of casting the heterogeneity of the contents of the receptacle (52e2) in terms of the comparison of the receptacle to various shaking instruments (52e7).

[42] The other component of the flux is analyzed at 58a–c. The two components are not to be distinguished temporally as the reference from 58c to 52e2 shows. There is no time when the phenomena just sift according to kinds and do not also undergo the processes described at 58a–c.

[43] The grouping of phenomena according to kinds whenever it is mentioned in the *Timaeus* is always described as the *result* of the moving of the contents of the receptacle, not as the *cause* of the moving of the contents (e.g., 57c). Cornford repeatedly mistakes this grouping according to kinds as the result of an active attraction of like kind to like kind. Cornford (2) 199, 202, 208, 228, 239, *et passim,* against which see Cherniss (1) n393, and Skemp (1) 58–59. The statement that there is no similarity of powers anywhere in the pre-cosmic state of the receptacle (52e2–3) dispels a necessary condition for Cornford's interpretation of the cause of the sifting of kinds, namely, that there are large masses of each of the primary bodies which act as gravitational centers for the rest of the contents of the receptacle prior to any movement in the receptacle.

Nor is the grouping of natural kinds to be confused with the grouping of natural kinds in Aristotle's cosmology. In the *Timaeus* regions of space are posterior to the contents of space. A fire particle moves because of properties it has as a fire particle and the region it occupies is fiery because it occupies it. For Plato it is not the case that a particle is potentially fiery and then becomes actually so as it enters into the natural fire region.

The interface of different kinds, as they pass between each other, does not result in friction or impediment, but, as we shall see later, results in the transformation of the kinds into each other. The interface results not in the cessation of motion, but rather the making chaotic the regular rectilinear motions that result from the relative buoyancies of different phenomenal kinds.

There are two backward references in the *Timaeus* to the description of chaos that cohere with and corroborate my interpretation of it. First, 58a1–2 seems to be a backward reference to 52e2.[44] The heterogeneity described in both passages seems to be viewed as a sufficient condition for motion at 57e7–58a1. Vlastos has correctly seen that this is the case,[45] though he does not argue for his interpretation of 57e7–58a1. Cherniss has challenged this interpretation, claiming that heterogeneity of kinds of particles is a necessary but not a sufficient condition for motion.[46] Cherniss' claim, though, is not advanced as having been gleaned from 57e7–58a1, but is advanced as a consequence of his overall interpretation of 52d–58c.[47] It seems reasonable, however, that since the expression οὕτω δὴ στάσιν μὲν ἐν ὁμαλότητι . . . ἀεὶ τιθῶμεν (57e6–58a1) means not that homogeneity is a necessary but that it is a sufficient condition for rest, the parallel phrasing κίνησιν δὲ εἰς ἀνωμαλότητα ἀεὶ τιθῶμεν (57e7–58a1) ought to mean that heterogeneity is a sufficient condition for motion.[48]

This is borne out by the ensuing lines (58a2–4) which are a second reference back to the description of chaos (53a). Importantly the passage 58a–c *presupposes* the analysis of motion given at 52d–53a. The passage does not try to explain how the motion there described comes about. That must have already been stated. This confirms the view, then, that the phenomena move spontaneously as they sort themselves according to kind. What the passage 58a–c contributes to the earlier passage 52d–53a is that, though the tendency of the primary bodies to separate into similar aggregates, or what I call natural buoyancy, is an active force which moves the primary bodies, the complete separation of all the contents of the receptacle as described at 53a2, 6–7 never actually occurs

---

[44] Just as 57e1–2 seems to refer to the whole of 52d–53a.

[45] Vlastos (2) 395–96.

[46] Cherniss (1) 448–49.

[47] Contrast Cherniss (1) 445, where whether heterogeneity is a sufficient condition for motion is left an open question, with (1) 449, where the question is answered in the negative.

[48] Easterling, because he fails to recognize that 58a1–2 and 57e1–2 refer back respectively to 52e2 and 52d–53a, mistakenly claims that 57e7–58a1 is not a discussion of either necessary or sufficient conditions for change. Easterling 38n29, against which see Tarán, n131.

(58a2–4). The description of the formation of ordered bands of primary bodies is not to be read literally. If such a separation actually occurred at any time, not only would there be no chaotic flux of phenomena, there would be no motion at all. For according to the principles enunciated at 57e such homogeneous bands would make motion impossible.[49] This means that some other principle must operate in order for the flux of the phenomena to be maintained. This principle is articulated in the rest of 58a–c.

II. MOTION AND REST: 57c–58c

We are now in a position to give the correct analysis of 57c–58c, by the end of which the mechanism is finally fully stated which "provides that the perpetual motion of the primary bodies is and shall be without cessation" (58c3–4). Cherniss and Tarán ultimately rest on this passage their analysis of the source of the phenomenal flux. I hope to refute their interpretation of the passage and establish in the process my own interpretation.

Cherniss holds that the flux of the phenomena arises when, in imposing order onto the static plenum of reflections of Ideas, the Demiurge accidentally but necessarily displaces phenomena in directions unrelated to his intentions.[50] The accidentally displaced phenomena necessarily in

---

[49] Note that the principles enunciated at 57e (referring back to 57a) specifically rule out the notion that any psychic intervention is here being entertained, since even in the case of homogeneity of particles, though the natural buoyancy which causes motion could not operate, psychic interventions could still be sources of motion.

[50] Cherniss does not give any textual citations for his view that the phenomena, just as reflections of Ideas in space, would be static. Cherniss (6) 255, (1) 454, on which see Vlastos, (3) 41n1. Cherniss simply appeals to the assumption that "according to Plato . . . nothing spatial or corporeal can be the cause of its own motion." Cherniss (6) 255. This though simply begs the question as far as the *Timaeus* is concerned. Another argument which Cherniss seems to use to try to establish that the phenomena just as reflections are static runs as follows. Premise one: "since the spatial mirror is homogeneous and the Ideas themselves are non-spatial, the reflections in space would not be locally distinct." Cherniss (6) n18. Premise two: spatial distinction is a consequence of demiurgic activity. Cherniss (6) n18. Premise three (implicit): locomotion entails that moving objects are locally distinct. Conclusion: there is no locomotion in the pre-cosmos. However, premises one and two are undermined by claims in the *Timaeus* A) that spatial distinctness arises simply from the presence of the reflections of Ideas in space (51b4) and B) that such spatial distinctness is indeed present in the pre-cosmos (53a2, 6). Further, though space itself is homogeneous, its contents are explicitly

turn move other phenomena. The Demiurge tries to appropriate for his purposes as many of these disorderly motions as he can, but in doing so he sets off still other accidental motions, so that the flux of the phenomena is constantly maintained.[51] This general view of the cause of the phenomenal flux has been attacked largely on methodological grounds,[52] not the least of which is that it is a view that Plato nowhere explicitly or implicitly states. No one, however, has probed Cherniss' interpretation of 57d–58c, in which he takes his view to be "exemplified rather than contradicted."[53]

---

stated not to be homogeneous in the pre-cosmos (52e2–3) and so must be locally distinct. Therefore, the conclusion that there is no motion in the pre-cosmos based on the above argument is false.

[51] Cherniss, (6) 258, (1) 448–50 with notes. This view has been defended, largely by mere repetition, by Tarán. Tarán 385–88 with notes.

[52] For references see n10 above. I should like to subtract one argument from and add one argument to the list of such general objections. First I do not agree with one argument which Herter raises against Cherniss. Herter points out that even if, as Cherniss asserts, the evil done by the Demiurge is accidental, the Demiurge would, on criteria from the *Laws*, still be subject to severe punishments, and since this is unthinkable, Herter claims the Demiurge must not be the cause, even the accidental cause, of erratic motions. Herter (1) 337–38. Herter, however, fails to distinguish between accidents (unintended actions) which result from negligence, and those, like the Demiurge's, which occur under the duress of Necessity. See *Laws* 901c for Plato making this distinction between acting negligently and acting under duress. The distinction is specifically mentioned as holding for divinity.

Against Cherniss though it should be noted that on his interpretation of the cause of chaos, the Demiurge ought to act like a utilitarian who interprets utility as order. Since the Demiurge knows that any intervention into the phenomenal world will produce chaos as well as order, he would need to calculate whether or not any one of his interventions would actually maximize order given that he will not be able fully to appropriate for his purposes all the chaotic motions which he will inadvertently but necessarily produce. On Cherniss' model, the Demiurge is rather like a sculptor who has been handed a marble block which he knows or discovers to be flawed. The sculptor has to decide whether order will be maximized by proceeding to work on the block, knowing that his intervention may shatter the block, or whether order will be maximized by simply abandoning the project. He may, of course, include in his calculation the knowledge that even if the block is shattered into wholly disordered rubble, the rubble can always be appropriated to some useful purpose by being burnt down into lime. But some utilitarian decision must be made given that there is a residue of disorder that results from such craftsman-like interventions. Yet in the *Timaeus* we never see the Demiurge making such order-utilitarian decisions which take into account the disorder which he will accidentally but necessarily work on the phenomena.

[53] Cherniss (1) 448.

Citing 57d–58c, Cherniss claims: "As the *Timaeus* explains, it ['the general flux of phenomena'] is the complex of secondary motions produced incidentally by the perfectly rational World-Soul as it induces directly the rational motion of rotation in the spherical plenum of spatial figures."[54] To support his view that the Demiurge or the demiurgic activity of the World-Soul is present in 57d–58c, despite the lack of any explicit reference to such activity, Cherniss takes the backward looking reference at 57e1–2 to refer to the whole of what has preceded in the *Timaeus* and in particular to the discussion of the revolution of the World-Soul.[55] However, since the immediately ensuing lines (57e2–6), which attribute motion to a condition of heterogeneity, are a summary and generalization, as Cherniss realizes,[56] of 57a3–5 and since the cause of heterogeneity is said to have already been described (58a1–2), a reference to 52e2, as Cherniss again concedes,[57] it is probable that 57e1–2 refers back only to 52d ff. and not to the discussion of the revolution of the World-Soul.

After referring at 58a1–2 to the cause of heterogeneity and so the motion that results from it (52d–53a), Timaeus claims that it has not been explained why the various primary bodies (earth, air, etc.) have not been completely separated apart into their kinds such that they cease to pass through each other and to change their places (58a2–4). The purpose of the rest of the passage (to 58c4) is to offer such an explanation (58a4). It is Cherniss' failure to see that this alone is the purpose of 58a–c which decisively undermines in several ways his interpretation of the cause of the phenomenal flux.

First, lines 58a2–4 presuppose the actualized tendency of particles to separate according to kinds. The passage 58a–c, then, is not, as Cherniss makes it out to be, an explanation of how it is possible that the phenomena are in such a condition that they do separate into kinds.

Second, the misunderstanding of the explicit purpose of 58a–c undermines Cherniss' reading of περίοδος (58a5). It is a necessary condition for Cherniss' interpretation that περίοδος here means "revolution" and not "circumference," such that it is a veiled[58] allusion to the demiurgic workings of the World-Soul and in particular to the revolution of the phenomena produced by the circuit of the World-Soul. That περίοδος (58a5) means "revolution" and not "circumference" Cherniss claims "is *proved*

[54] Cherniss (6) 258.

[55] Cherniss (1) 449, 450; also Tarán n115.

[56] Cherniss (1) 444–45 and n383.

[57] Cherniss (1) 444.

[58] Note that allusions to demiurgic activity in the discussion of Necessity (47e–69a) tend to be quite explicit (47e–48a, 53a–b, 56c, 68e–69a). This would suggest that a "veiled" reference to demiurgic activity is no reference at all.

by" the earlier occurrence of the term at 34a6.[59] This earlier usage, however, *proves* nothing of the sort. If the term "carat" appeared in a jewelry advertisement, and if the first occurrence referred to a unit of weight and not to degree of purity, we would be quite mistaken to suppose that the second occurrence had also to refer to weight. Whether it referred to weight or to degree of purity would depend upon whether it was used in describing gems or gold, and this would apply quite independently of how the term was used in its first occurrence. Which sense of περίοδος is intended at 58a5, then, has to be determined by context and the context is governed by the purpose stated at 58a2–4. To fulfill this purpose, περίοδος in the sense of "circumference" is required. For the role the περίοδος fulfills in the purpose stated at 58a2–4 is to keep the contents of space confined, so that the contents are kept constantly closely packed. If the universe were capable of expansion, the smaller particles, in escaping (ἐκφύγῃ, 57b6) from their interactions with larger particles, could avoid being thrust into the interstices between the larger and therefore necessarily less efficiently packed particles (58b4–5) (such that heterogeneous contacts would be maintained) and so could escape into open outer spaces and form a homogeneous band, which according to the principles laid out at 57e would bring the motion of the particles to a halt. It is the finitude of the universe that is the only relevant concern of the passage. Περίοδος at 58a5, then, means "circumference." As Cornford points out, if περίοδος (58a5) referred to a revolution of the whole universe, there would not be a tendency for the universe to remain finite but for it to expand by centrifugal force.[60]

Those, like Taylor, who do take περίοδος to mean "circumference" take it to refer to the circumference of the formed οὐρανός.[61] This has lead Tarán to claim that even if περίοδος (58a5) meant only "circumference" and not "revolution," then since the περίοδος is said to be round, it is implied that demiurgic action is still necessary for the maintenance of chaotic motion.[62] However, I do not think that περίοδος (58a5) refers to the οὐρανός, the formed contents of space, but rather refers to space itself. For it is space and not the formed universe that is said to be full, simply as the result of the Ideas being imaged in it (52d–e). Since this imaging occurs over the whole extent of space (51a2), the only way the contents of the receptacle will remain confined and thus fulfill the stated

---

[59] Cherniss (1) n392, my emphasis. Tarán also claims that *Timaeus* 34a6–7 "*proves*" περίοδος at 58a5 means "revolution." Tarán n122, my emphasis.

[60] Cornford (2) 243–44.

[61] Taylor (1) 398.

[62] Tarán n126. So too Cornford, (2) 243: "Spherical shape, then, including all the bodily, is a new factor here introduced, as a work of Reason . . . ."

purpose of 58a–c is if the receptacle itself is circumscribed. So the need of maintaining a finite extension applies to space, and only indirectly to its contents.[63] Περίοδος (58a5), I suggest, then, refers to the circumference of space.

If this is so, Plato specifically tells us that the shape of space is round (κυκλοτερής, 58a5).[64] There is no explanation of why space is round, but being round it keeps the particles from getting too far away from one another. If space were irregularly shaped, there could be coves into which homogeneous particles might drift and thus fall outside of the general flux of phenomena. Space is also said to come together upon itself (πρὸς αὑτὴν συνιέναι, 58a6). I take this to mean that the figure space describes is closed, unlike, say, a spiral or cornucopia. This is a consequence of its being round[65] and is an additional reason why space keeps particles from wandering off to infinity. No rational, demiurgic, or psychic activities are entailed, then, by lines 58a4–7.[66]

Nevertheless, Cherniss, in assuming that περίοδος at 58a5 refers to the revolution of the World-Body caused by the revolution of the World-Soul, claims this rotary motion of the World-Body causes the various particles, as a result of their different sizes and configurations, to "jostle one another."[67] This jostling is equated with the flux of the phenomena, and, in turn, according to Cherniss, is the cause of the separation of the phenomena according to kinds.[68]

Cornford, at the very end of his discussion of motion and rest, almost as an afterthought, entertains an explanation of the cause of the phenom-

---

[63] That the formed universe has to be co-extensive with the whole of space, and therefore that space is round can be deduced on independent grounds. If the formed universe were smaller in any way than space, the rest of space would fill up with disorganized images (51a2 with 52d–e) and the universe would no longer comprise all the elements of chaos (30a3–5). It is apparently from this line of reasoning that Cornford correctly concludes that since the universe is round, so too must be space. Cornford (2) 188. That the shape of space is round can be deduced on grounds independent of 58a4–6 torpedoes the argument of those who claim that the rounded περίοδος of 58a5 must refer to the product of rational activity.

[64] Cornford correctly points out that κυκλοτερής "is more appropriate to shape than to movement, and is applied to shape at 33b." Cornford (2) 243.

[65] The καί of 58a6, then, means "and so."

[66] With Taylor, Cornford, and Cherniss, and against Skemp, I do not take σφίγγει (58a7) as implying any "inward pressure" or "constrictive force," but rather as meaning "encompassing round about." Taylor, (1) 397–98, Cornford, (2) 244, Cherniss, (1) n392, Skemp, (1) 63.

[67] Cherniss (1) 449 with n393.

[68] Cherniss (6) 258, (1) n393.

enal flux which is similar to that given by Cherniss.[69] Cornford claims that "the order of the layers [of primary bodies as entertained at 53a] could be explained as due to the rotary movement (a work of Reason), sifting the more mobile particles toward the circumference, the less mobile toward the center, on the familiar analogy of an eddy in water collecting the heavier floating objects at its center."[70] Cornford' s model here and Cherniss' model are similar in that both view the flux of phenomena ultimately as an indirect effect of the rational demiurgic activities of the World-Soul which are manifested in the rotation of the universe. The two models differ, however, as follows. For Cornford the rotation of the phenomenal plenum is the *direct* cause of the separation of kinds, and this separation is in turn the *direct* cause of the flux of phenomena. For Cornford, then, the rotation of the plenum is the *indirect* cause of the phenomenal flux. Just the opposite holds for Cherniss, who takes the rotation of the plenum as the *direct* cause of the phenomenal flux, which in turn is the *direct* cause of the separation of like kinds. So for Cherniss, the rotation is the *indirect* cause of the separation of like kinds.

To appreciate the difference between the two models imagine a revolving disk with but one particle on it. On Cornford's model, if the particle were heavy, it would move considerably more slowly than the rotation of the disk *and* would drift toward the center of the disk, like a phonograph needle which has accidentally been dropped on a moving turntable; if the particle were light, it would move more nearly at the speed of the disk *and* drift toward its edge. On Cherniss' model no matter what the weight of the particle it would remain on a single circular track, not moving either nearer or farther from the center. The heavier the particle, though, as on Cornford's model, the greater the reduction of its speed off the speed of the disk, such that if two particles of different

---

[69] Cornford (2) 246.

[70] Cornford (2) 246; cf. Cherniss (1) n393. The model for explaining the movement of the phenomena given here by Cornford stands in blatant contradiction to his two claims (themselves not obviously compatible)—1) that an irrational element in the World-Soul is the cause of the flux of the phenomena and 2) that an *attraction* of like to like is the cause of flux. For on this model, i) the flux occurs as the indirect effect of the rationality of the World-Soul rather than the direct effect of the irrationality of the World-Soul and ii) the *attraction* of like to like is apparent rather than real. See Cherniss (1) n393. This model of Cornford's is neither implicit nor explicit in the *Timaeus,* and Cornford does not cite any evidence from the *Timaeus* for it. Rather he resorts to citing Aristotle's claim that the comparison of the cosmos to an eddy was invoked by "all those who try to generate the heavens, to explain why earth came together at the center" (*De Caelo* 295a13). Cornford (2) 246. Cornford's model is subject to the same criticisms which I have raised to this point against Cherniss' model.

weight were on the same track, the lighter of the two would eventually overtake and collide with the heavier. It is this type of collision which Cherniss calls the jostling of the phenomena. This jostling which is directly caused by the revolution of the universe, on Cherniss' model, in turn, is the cause of the separation of phenomena according to kinds.[71] However, of this jostling that results from one particle overtaking another on the same track concentric to the rim of the universe there is not a single word in the *Timaeus.* Nor does Cherniss cite any line where he supposes that this is mentioned. It certainly is not mentioned at 58a as the result of ἡ τοῦ παντὸς περίοδος (58a4–5). The only result there mentioned is the tendency for there to be no room left empty (58a7). Further in the description of the transformations of the primary particles there are some claims made which, by referring back to the separation of kinds at 52a–53a, suggest that the collisions or gratings between different kinds of particles are being viewed as occurring not along circular tracks which are concentric to the circumference of the universe but along lines extending from the center to the circumference of the universe.

First, take the statement that the masses which envelop fire, which itself is moving (κινούμενον), are moving in place (φερομένοις) (56e4). On Cherniss' analysis, φερομένοις must refer here to the non-fire particles moving along the same circular track as the fire particles and being overtaken by them. However, at this point, even on Cherniss' analysis, we have heard nothing whatsoever that would make us suspect that this is the case, since the crucial occurrence of περίοδος appears only at 58a. On the contrary, description of the motion of fire and non-fire particles which results in their interaction and transformation (56e4) pretty clearly refers back to that motion in the description of chaos which is characterized as εἰς ἑτέραν . . . φερόμενα ἕδραν (53a2). The interactions of fire and non-fire particles occur, then, as the result of fire moving, relatively to the non-fire particles, toward the circumference of the universe and the non-fire particles moving, relatively to the fire particles, toward the center of the universe.

Second, as the result of being overcome and made similar to a non-fire mass, fire αὑτοῦ σύνοικον μείνῃ (57b7). On Cherniss' interpretation this would mean that the transformed fire particles take on the same speed in relation to the speed of the universe's rotation as the non-fire particles. But again there is nothing to lead us to anticipate such an interpretation, since in light of the immediately ensuing summary of 52e–53a at 57c, the expression αὑτοῦ σύνοικον μείνῃ would seem to refer back to the description of the separation according to kinds given already at 53a (n.b. ἵζει, 53a2), and so seems to be referring to a *position* relative to the center

[71] Cherniss (1) n393. For a discussion of the jostling, see below.

of the universe, rather than to a *speed* relative to that of the universe's rotation.

The description of the transformation of the primary particles, then, does not seem to recommend Cherniss' reading of the cause of the flux of the phenomena. Rather the cause of the transformation is the grating of particles that results from the inherent actualized tendency of the particles to separate according to kinds. The description of the transformations and of the consequences of the transformations conform to such an interpretation of the cause. So the flux of phenomena, which is equated with the continuous transformation of particles (58c), seems to arise as the result of the tendency of the phenomena to separate according to kind.

Cherniss asserts just the opposite, claiming that the sifting of the phenomena according to kinds is the *result* of the particles' jostling, which he equates with the phenomenal flux in general. He claims that, in consequence of different particles having different resistances to being set in motion (as stated at 55e–56a), "as the corpuscles jostle one another in the revolving plenum, those of earth would tend to settle at the center and next to them the corpuscles of water" and so on.[72] Cherniss does not spell out the mechanism by which the sifting according to kinds is to result from the jostling, but the transformations of particles that result from the jostling do not in any obvious way result in the sifting of kinds. For on Cherniss' model, the transformations of kinds would simply adjust the speed at which the particles move along a given circular track, but would not entail a change of track. If on the other hand, Cherniss supposes that some sort of carom-like rebounding of particles produces deviations from the original track of the particles, then, though the differences of degree of resistance to movement, which different types of particles have, might influence the force or distance of the particles' rebounds, they would not affect the direction of the rebounds toward or away from the center of the universe. So there would be no overall tendency (actualized or not) for particles to form separate regions according to kinds. Indeed a carom-like rebounding of particles, given an initial random distribution of kinds of particles (as is asserted at 52e2), would tend to maintain the heterogeneity of the contents of the receptacle rather than to create homogeneous layers.

Though Cherniss' general explanation of the drift of phenomena according to kinds fails, and though he does not elaborate the mechanism of the drift, he does cite two passages in which, he claims, the drift is indeed the *result* of the flux of the phenomena.[73] One is 53a, the initial de-

[72] Cherniss (1) n393.
[73] Cherniss (1) n393.

scription of the separation of phenomena according to kinds. I have shown above, though, that Cherniss is wrong in supposing that 53a suggests the view that the sifting of kinds is a result of the flux of the phenomena (section I).

The other passage which Cherniss takes as suggesting this view is 57b5–6. He translates ὠθούμενα καὶ διαλυθέντα (57b5–6) as "extruded and so liberated."[74] Cherniss cites *Laws* 904d3–4 for a parallel use of διαλύω meaning "liberated." But given the many other occurrences of the term in the immediate context (53e2, 54c4 and especially, 56d2, 58b1; cf. λύω, 54c6, 56d3, 57a6, 57b5), διαλυθέντα at 57b5–6 clearly refers to the dissolution of particles into their constituent triangles, which are then capable of being reconstituted into the same or different types of particles, and does not mean "liberated" as might be suggested by the accompanying ἐκφύγῃ (57b6). Rather, the lines 57b5–6 propose the view that in an interaction in which one type of particle is overcome by another, one possible result is that the particles which are overcome dissolve into their constituent triangles, which then can be reconstituted into their original kind (though elsewhere than in the area of the original interaction) such that 1) the area of the original interaction is left homogeneous and (2) the reconstituted particles begin again to move based on their natural buoyancy so they escape to their kindred. The lines 57b5–6, then, do not constitute a proof-text for the view that the grouping of similar phenomena is the result of the particles' jostling, which for Cherniss constitutes the phenomenal flux. I suggest then that Cherniss' attempt to show that the sifting of kinds is the result of the flux of the phenomena fails.

More generally, it is surprising that Cherniss does not find, and does not even claim there is to be found, an *explanation* of why the particles sift according to kinds, and that he is content with just the *assertion* that they do sift as the result of the flux of the phenomena. It is surprising because for Cherniss, when the original problem is raised of how Necessity causes motion, the motion under consideration is ultimately cashed out as the very drift of the primary particles, which is the product of the flux or Necessity.[75] It seems that on Cherniss' understanding of Necessity and his reading of 52d–58c, the original sought-for *explanation* of how Necessity causes motion is never forthcoming.

If, however, we view the tendency of the phenomena to separate according to kinds as the primeval efficient cause of motion in the universe, as has been suggested by my analysis of the description of chaos, then the sifting according to kinds fits into the economy of the dialogue and in particular jibes with the expressed purpose of 58a–c. For the sort-

[74] Cherniss (1) n393.

[75] Cherniss (1) n388.

ing of like kinds introduces an anomaly into the discussion of the phenomena which 58a–c resolves, thus saving the phenomena (58c3) and giving the full analysis of the phenomenal flux. If the separation of kinds were ever a fully actualized state of the universe, rather than just a tendency, all motion would cease, as each sector of space would then have a homogeneous content (58a). At 57a and 57e–58a, that homogeneity is a sufficient condition for rest is left an unanalyzed assertion. On Cherniss' account homogeneity is a sufficient condition for rest because, if all the particles on a circular track parallel to the circumference of the universe are of the same kind, all will move at the same speed and so none will ever overtake and jostle any other. Therefore, of the properties assigned to the particles at 55e–56b as the result of their geometrical shape, it is mobility or ease of being set in motion[76] which Cherniss sees as relevant to homogeneity resulting in rest.[77] On this model, though, the kinds would not need to have separated completely (58a3) into *four* bands of a determinate order (53a) for homogeneity, which results in rest, to occur. On Cherniss' model all that is required for the cessation of motion is that all the particles on *any* given track around the center of the universe be homogeneous. The universe could have any number of concentric homogeneous bands in any order on Cherniss' model and still be completely at rest according to the principles of 57e.[78] This problem suggests that mobility or at least mobility with respect to the rotation of the universe is not the property of the primary particles which is relevant to homogeneity resulting in rest. Rather, the only property of those mentioned at 55e–56b which can adequately account both for the number and order of bands in the hypothesized state of homogeneity (58a3) is weight (56b1–2).[79]

We need to distinguish two different senses of weight in the *Timaeus*. Weight, on the one hand, is treated in an objective sense as being an inherent property that the primary particles have as a direct result of their geometrical construction (56b2) and not as the result of any interaction between their construction and a perceiver.[80] Later, weight, on the other

---

76 That is, by the revolution of the universe on Cherniss' model.

77 Cherniss (1) n393.

78 This problem also besets Zeyl's account of homogeneity, which fails to account for the number and order of the hypothesized homogeneous bands. Zeyl (2) lxii.

79 On ἐλαφρότατον, see Cornford (2) 222n4 and Cherniss (1) n85.

80 See Cherniss (1) 138. Although weight is assigned according to the number of sides a particle has, this does not mean a side when it is drifting about, not a part of a particle, has a weight. Further, since it is the number of sides which determines the objective weight of a particle, size is not a determining factor in weight. All fire particles no matter what their size have the same objective weight.

hand, *is* taken in a subjective sense as a property which arises only between a perceiver and the stereometric construction of the primary particles (62c–63e).[81] Weight in this subjective sense, however, depends upon the natural tendency of the phenomena to sift into concentric bands (63e).[82] It is weight taken objectively that adequately explains why homogeneity results in rest. The various types of particles have an objective weight in relation to each other as the result of their configurations (56b1–3). Thus when a particle of less weight is nearer the center of the universe than a heavier particle, there is a tendency for them to exchange places. This accounts, then, for both the number and order of bands that would arise if the separation of kinds were ever fully actualized. For the interchanging of heavier with lighter particles, as described above, 1) would not allow a band consisting of a lighter type of particle to be nearer the center of the universe than one of a heavier type and 2) would cease only when all particles of the same type were together. Together 1) and 2) would result in an arrangement of four concentric bands with the band of the heaviest type of particle at the center and the lightest at the rim of the universe. The reason homogeneity of a kind within a region of space results in rest is that since all particles of the same type have the same objective weight,[83] there would be no tendency for two particles of the same type to interchange regions.

It seems then that objective weight, by accounting for the separation of kinds,[84] is being treated as the ultimate cause of the physical changes which constitute the phenomenal flux. When, then, Plato writes that we must always attribute motion to a condition that is heterogeneous (57e7–58a1), he is stating, as we suggested earlier (section I) a sufficient condition for change and not just a necessary condition for change. As we

---

[81] The analysis of weight at 56b1–2 and the analysis of weight at 62c–63e are in blatant contradiction to each other if they are taken as describing the same aspect of the phenomena. Cherniss tries to get around the contradiction by claiming that the later analysis supersedes and negates the earlier analysis. Cherniss (1) n85. This way out of the contradiction, however, leaves Plato looking pretty silly and forgetful, since Cherniss cannot find any reason, even a heuristic reason, for the earlier analysis and, on Cherniss' view, the later analysis is not a clarification or development of the earlier analysis. Rather one should recognize that there is no contradiction between the two analyses. They do not describe the same aspect of the phenomena.

[82] So that if one carries a mass in the direction in which it naturally drifts (say, fire toward the rim of the universe) it *feels* light. But, if one carries a mass in the direction opposite to its natural drift (say, fire toward the center of the universe), it *feels* heavy.

[83] See n80.

[84] Therefore, since weight *as* perceived depends on the tendency of the kinds to separate (63e), subjective weight presupposes objective weight.

were led to expect by the opening lines of the description of chaos (52d4–e3), the cause of the separation of kinds and the flux of phenomena which results from the separation (see below) is the result of properties which the phenomena have just as being images reflected in the mirror of space.

(An Analysis of Timaeus 58a–c)

Given the tendency for the primary particles to separate according to kinds—a tendency which results from their natural buoyancy, which, in turn, results from their objective weight—the establishment and maintenance of the phenomenal flux can be explained without appeal to either demiurgic organization or psychically induced motion. Rather given the inherent actualized tendency of the phenomena to separate according to kinds, the sufficient conditions for the establishment and maintenance of the phenomenal flux are 1) the nature of space and 2) the geometric properties of the primary particles. The relevant property of space is that it describes a regular closed figure (58a5–6), so that given that images are cast over its whole extent (51a), it tends to allow no room to be left empty (58a4–7). The relevant properties of the primary particles are a) that the larger particles are the less efficiently they can pack together: "The kinds that are composed of the largest particles leave the largest gaps in their texture, while the smallest bodies leave the least" (58b2–4) and b) that the process of condensation (fire changing to air, air to water) results in the same number of triangles enclosing an increased volume (58b4–5).[85] Because of a) and b), when, as the result of the grating of particles,[86] smaller particles are overcome by larger particles such that some of the smaller particles form into larger particles, there is an overall increase in the contained volume in the area of the interaction, and since space is finite and does not expand, there is a pressure set up in the area of the interaction.

Some critics suppose that this pressure of expansion is relieved by a compensatory decrease in volume through rarefaction (water changing to air, air to fire) somewhere else in the universe.[87] But we hear not a word in the *Timaeus* of such a cycle of compensation. It is more likely that the pressure created locally by condensation is relieved locally by an increased efficiency of packing. And this indeed is what is said to occur.

---

[85] The volume of one water octahedron is larger than the combined volumes of two fire pyramids (given that the faces of the pyramids and octahedron are triangles of the same grade). See Cornford (2) 245 with n2 and Vlastos (8) Table I, 89.

[86] The grating arises as the result of i) the particles moving due to their natural buoyancy and ii) their being closely packed.

[87] Cornford (2) 246, Vlastos (8) 90 with n35.

The smaller particles which have not been crushed in the interaction with the larger particles or which have been dissolved but have reformed into their original kind are said to be thrust into the interstices between the larger particles (58a7–b2, a passage stated as a conclusion [διὸ, a7] from 58a4–7 *and* 58b2–5 [note γάρ, b2]). By their natural buoyancy, the smaller particles would tend to drift toward the rim of the universe, but since this area is already closely packed as the result of i) smaller particles' tendency to occupy the outer reaches of the universe and ii) smaller particles packing more efficiently than larger ones, the local pressure cannot be relieved by the natural drift of the particles. Rather the pressure is relieved by the thrusting of the smaller particles into the interstices between the larger ones with the result that I) the smaller particles are thrust in a direction opposite to the direction in which they travel as the result of their natural buoyancy (58a7–b2) and so (οὖν, 58b6) II) larger and smaller particles, i.e., particles of different types, are kept in constant contact with each other, and so in turn (οὕτω δή, 58c2) III) heterogeneity is constantly maintained throughout space.

So in general, the natural buoyancy of the primary particles *directly* causes different types of particles to separate according to kinds, but as a result of the properties of space and of the primary particles, it *indirectly* causes the particles to be thrust back together again, and so the flux of the phenomena is caused and maintained. The "counter-force" which keeps the different particles thrust together as the indirect consequence of the natural buoyancy of the primary particles has no psychic or demiurgic cause whether direct or indirect. The flux of the phenomena, then, according to the *Timaeus* arises from and is maintained by purely mechanical non-psychic causes. In the *Timaeus* the phenomena are, on their own, a spontaneous source of disorder and positive evil.

This is, I suggest, as the reader of the introduction to the *Timaeus* (17a–27b) would expect it to be. Here Critias advances the Atlantis story as an allegory (26c–d) by which he can portray the perfect government in action. The role Timaeus' discourse is to play in Critias' plan is not explicitly stated (cf. 27a). However, its purpose pretty clearly is to elaborate on that which in the allegory is represented by natural disasters (conflagration, flood, earthquake).[88] In the Atlantis story such natural disasters are set over against the actions of moral agents. They are portrayed as operating in opposition to and independently of any moral agent (25c–d). In the allegory the Atlantians represent moral evil (25c); their evil arises from ignorance, which in turn arises from forgetfulness (*Critias* 121a–b). The Athenians represent moral excellence, which arises from wisdom (*Timaeus* 24b–c). The natural disaster which indiscriminately destroys

[88] So too, Broadie (1) 2–4.

both the Atlantians and the Athenians, and which completely wrecks the best demiurgic accomplishments of divinity (24c4–5) is repeatedly portrayed as a positive, non-moral source of evil which arises spontaneously. The introduction of the *Timaeus,* then, sets up the reader's expectation for an account of such a positive evil which arises spontaneously from the sheer corporeality of the phenomena. The interpretation which I have given of Timaeus' discourse satisfies this expectation.

PART THREE

THE OTHER COSMOLOGICAL WRITINGS

SEVEN

Disorderly Motion in the *Statesman**

In this chapter I will argue that in the *Statesman* myth's account of alternating cosmic cycles, Plato views the phenomena, or physical objects, in and of themselves as a positive source of evil, moving erratically without psychic causes whether rational or irrational, direct or indirect.[1] In this respect and others, the cosmological commitments of the *Statesman* are the same as those of the *Timaeus*[2] and are inconsistent with those of the *Phaedrus* and *Laws* X, which view souls as self-moving motions, which in turn are the sources of all other motions (*Phaedrus* 245c9, *Laws* 896b1).[3] In the *Statesman* and *Timaeus,* Plato views the soul, as in the

---

* Reprinted from *Phoenix* 35 (1981) by permission of the editor.

[1] The view that the disorderly motions in the *Statesman* have no psychic causes has also, but for reasons different than the ones I will advance, been held by Vlastos and Herter. Vlastos (2) especially 394–95, Herter (2) especially 111.

[2] See chapter 6 above.

[3] Robinson, (1) 134–35, 139, mistakenly, I think, takes the *Statesman* myth to be a synthesis of the autokinetic doctrine from the *Phaedrus* and the cosmology of the *Timaeus,* which he dates earlier than both the *Statesman* and *Phaedrus* and in which he (correctly, I think) finds no trace of the autokinetic doctrine. Robinson takes several reflexive phrases in the myth as referring to autokinetic soul:

> (i) αὐτὸ ἑαυτὸ στρέφειν ἀεί, 269e5 (of the Demiurge). But στρέφειν cannot merely be a synecdoche for κινεῖν, for so construed it does not fulfill the demands of the context. It is because the Demiurge moves constantly *in one direction* rather than because it is autokinetic that it is said that the Demiurge cannot cause two contrary motions. That the phrase is reflexive merely means that the Demiurge's rotation is independent and non-contingent, in contrast to the rotations of the world.
>
> (ii) τῆς αὐτοῦ κινήσεως, 269e4. The antecedent of αὐτοῦ, though, is the World-Body (269d7–8), not the World-Soul, which it comes to possess (d8–9, with d1). The phrase merely describes the motion of the World-Body (or World-Body/World-Soul complex) as it is moved in the train of the Demiurge's rotation. The term αὑτοῦ means something like "proper to itself under the best of conditions."
>
> (iii) finally, δι' ἑαυτοῦ, after Burnet, 270a5. This expression may simply be taken mechanistically (as Robinson admits it might be), for the immediately ensuing account of the world's reverse motion is *described* entirely on a mechanistic model (270a6–8), even if one wishes to claim Plato *means* something else. Thus the phrase ceases to be direct evidence for autokinesis. But in any case, I think the reading of the BT manuscripts should be preserved:

Phaedo and *Republic* X, as a substance rather than as a kind of motion. Further, in the *Statesman* and *Timaeus*, the World-Soul, which extends throughout the whole World-Body and which is paradigmatic for souls in general (*Timaeus* 41d), is not even viewed as a cause of motion, but instead as a maintainer of orderly motion against the spontaneous chaotic motions of the corporeal,[4] while Plato's unique, divine, demiurgic Reason or God, in these dialogues, though he is a source of motion—that of all orderly motions of both bodies and souls—is not himself a soul.[5] It would seem that on the nature of God, soul, and the corporeal there will be no way of reconciling the *Statesman* with the *Phaedrus* and *Laws* X, if it turns out that physical motions in the *Statesman* myth are spontaneous rather than psychically induced.

Many critics have tried to square the presence in the *Statesman* myth of motions which on the face of it have no psychic causes with the autokinetic doctrine. The strategies critics have used to bring about a reconciliation are the same as have been attempted to square the *Timaeus* with the autokinetic doctrine. There are four such strategies. One view is that an irrational part of the World-Soul is the cause of erratic motions.[6] As far as the *Statesman* is concerned it is sufficient to point out, against this view, that what irrationality the World-Soul has *is caused by* rather than causes disorderly physical motion (273a, c–d). A second strategy, having its roots in Plutarch, is that there is an evil counter World-Soul which works over and against the rational World-Soul and which is the source of chaotic motions.[7] This view runs up against the explicit denial of a Zoroastrian model of explanation for celestial dynamics (at 270a1–2).[8] A third reconciliatory strategy is to claim that disorderly motions are inadvertent but inevitable spin-off effects from the orderly actions of the wholly rational World-Soul.[9] As an interpretation of a text rather than as an attempt to save the text from itself, this view has to read the mechanical model at 270a6–8 as the explanation of the efficient cause of the reverse circuit of the universe. But the model is meant only as an explana-

δι' ἑαυτόν, meaning "throughout itself" and looking forward to σεισμὸν πολὺν ἐν ἑαυτῷ, 273a3. In this case the phrase carries no causal or instrumental force.

4 See chapter 9.

5 See chapter 10.

6 This view is held by Cornford (2) 57, 175–77, 205–206, 209–10, and Morrow 437.

7 Plutarch, *On the Psychogony in the Timaeus* in C. Herbert (ed.), *Moralia* 6.1 (Leipzig: Teubner, 1954) 1014a–1015f. This view was advanced in the twentieth century by Skemp (1) 74–78, 82–84, 112, Dodds 115–16, and Clegg.

8 See chapter 6, n7.

9 This view is held by Cherniss (1) 444–50, (10) 458–61, (6) (on the *Statesman* myth in particular, see nn21, 44) and by Tarán 386–88.

tion of why the reverse circuit lasts as long as it does; its efficient cause is assumed *already* to have been explained (270a7–8, note especially διά, a7). A fourth strategy tries to claim that the autokinetic doctrine is intended by Plato only to explain the source of motions in the orderly and ensouled cosmos and is not meant to explain the source of motions in pre-cosmic or acosmic eras.[10] The special difficulty which the *Statesman* myth presents this strategy is that there still must be found psychic causes for the disorderly motions of the retrograde world, since it, though disorderly, is still an ensouled world. This would involve abandoning the strategy or reverting to one of the earlier alternatives.

Now, I take it that the cause of disorderly motion is simply equivalent to the efficient cause of the retrograde cycle. This is obviously the case if the alternating cycles are intended to be read non-literally, that is, as dramatic representations of constitutive factors which in fact obtain simultaneously in the actual world.[11] For then the reverse circuit represents nothing over and above the disorderly motions in it. But even if we do suppose the system of counter-cycles is to be read literally, it is important to notice that the initial motions of the world, after the withdrawal of the Demiurge, are chaotic. That these motions become (at least temporarily) ordered again during the counter-cycle into a circular revolution is the work of the World-Soul remembering the teachings of the Demiurge (273b1–2). The World-Soul is, I suggest, not being viewed primarily, if at all, as a source of motion but rather as a maintainer of order against the naturally inherent tendency of the corporeal towards disorder. Plato, I suggest, is distinguishing the *cause* of the reverse circuit *qua* motion from the *cause* of the reverse circuit being orderly and circular. Therefore the cause of the reverse circuit is equivalent to the cause of disorderly motion. This cause which we are looking for is a proximate efficient cause; there is no suggestion that the reverse circuit is simply a ricochet effect.

I will suggest that it is the purpose of the argument at 269d5–270a2 to explain the proximate efficient cause of the reverse circuit. Once the argument is parsed, it becomes fairly easy to sort out what in the rest of the myth is to be taken literally and what figuratively.

The argument (B below) with its immediate context (A and C below) runs as follows (269c4–270a8):

A At one time God himself guides and transports this world in its revolving course, but at another time, when the revolutions have at

[10] For a critique of this now widely held view, see chapter 8.

[11] The question of the meaning of the alternation of cosmic cycles can and should be isolated from the question of whether Plato believed in an *initial* act of world formation. See chapter 13.

length reached a measure of time allotted to him, he lets it go, and then, in turn, the world, which is (ὄν) a living creature and is allotted intelligence by him who fashioned it in the beginning,[12] revolves on its own (αὐτόματον) in the opposite direction. Now this reverse motion is a necessarily inherent part of its nature (ἐξ ἀνάγκης ἔμφυτον) for the following reason (διὰ τόδ'):

B I To stand always in accordance with the same things and similarly and to be always the same (τὸ κατὰ ταὐτὰ καὶ ὡσαύτως ἔχειν ἀεὶ καὶ ταὐτὸν εἶναι) is a condition fitting only to the most divine of all things. But, the nature of a body is not of this disposition. And what we call the heavens or the formed universe, while, on the one hand, it has received many blessings from its parent, nevertheless, on the other hand, also partakes in a body. Therefore, it is impossible for it to be free from change throughout its entirety.

B a Nevertheless (γε μήν) to the extent that it is possible the world moves in a single uniform course in the same place (ἐν τῷ αὐτῷ κατὰ ταὐτα μίαν φοράν). Therefore, it is allotted a countercycle which is the least deviation from its own (αὑτοῦ) motion.

B b i Now, to turn itself always (αὐτο ἑαυτὸ στρέφειν ἀεί) is hardly possible for anything but him who guides all the things which move backwards (αὖ).

B b ii And (δέ), it is contrary to divine law (οὐ θέμις) for him to move now in one direction and now in the opposite direction.

B II Now as the result of all this (i.e., Bb and the facts in A), we must not say either that the ordered universe turns itself always,

B III or again that it is always turned by God in two opposite courses,

B IV or again that two gods with opposed thoughts turn it.

C The only remaining alternative is what was earlier stated (ὅπερ ἄρτι ἐρρήθη καὶ μόνον λοιπόν) (namely BI), that at one time it is guided by an external divine cause, and gains the power of living again and so receives a restored immortality from the Demiurge, but at another time when it is let go, it moves (itself, ms B) throughout itself (δι' ἑαυτόν, mss BT), being left to itself in due season (as indicated previously) such that it carries itself backwards through many myriads of circuits, because it is very large and most evenly balanced and proceeds upon a very small foot.

The overall structure of the argument is the following. Section A establishes what it is that the argument is trying to prove. The argument proper, B, is cast as a large-scale disjunctive syllogism. We are offered what is taken to be an exhaustive list of alternatives (BI, II, III, IV). Then, one by one, alternatives are eliminated until but one alternative remains (BI). It is, then, asserted to be the case by the force of the syllogism. The

---

[12] On the significance of formation κατ' ἀρχάς in the myth, see chapter 13.

start of the argument is signaled by the διὰ τόδ' of 269d2, and the phrase ὅπερ ἄρτι ἐρρήθη καὶ μόνον λοιπόν of 270a2 both indicates the form of the argument as a disjunctive syllogism and ends the list of possible disjuncts. Section C, then, is an elaboration of the disjunct that is asserted by the force of the syllogism. The asserted disjunct is BI. Section C is not to be viewed on its own as the point of what is established by the argument, but must be interpreted in light of BI. Nor is section C merely a restatement of the facts contained in A, despite a strong similarity in language between the two passages. Section Ba is a promissory note, claiming that in fact it is going to turn out that the reverse motion established in BI is of a uniform sort; the note is made good only much later at 273a–e. Section Bb contains extra information, over and above that given in section A, which is used in eliminating the unsuccessful disjuncts (BII, III, IV). The content of the argument is analyzed below.

SECTION A

The purpose of section A is to show that the argument B will explain the proximate efficient cause of the reverse circuit of the universe. Many translators (Skemp, Fowler, Taylor) and critics take the issue of the cause of the reverse motion of the universe as already settled within section A itself by reading the participial construction ζῷον ὂν καὶ φρόνησιν εἰληχός . . . (269d1–2) causally rather than merely descriptively. Such a reading presupposes some rational psychic or animating force, namely, the rational World-Soul, as the cause of the reverse circuit and as an explanation of the expression αὐτόματον (269c7). Two points, however, weigh against this reading. First, it seems to be contradicted by the discussion of causes at the end of the *Sophist* where spontaneous causes (αἰτίας αὐτομάτης, 265c7) are said to generate without intelligence (ἄνευ διανοίας φυούσης, 265c8) and are specifically contrasted to causes, like the World-Soul, which arise from Divinity (ἀπὸ θεοῦ, 265c9) and are endowed with reason (μετὰ λόγου τε καὶ ἐπιστήμης, 265c8). It is left open whether these spontaneous causes are irrational psychic causes or nonrational mechanistic forces.[13] But rational psychic forces seem to be ruled out. Second, whatever is being explained by argument B is being ex-

---

[13] Skemp clearly is going beyond the text when he claims that only irrational psychic forces are meant. Skemp (1) 21. Further, insofar as the "commonly expressed opinions" which Plato is rejecting out of hand at *Sophist* 265c look very much like the views of Empedocles, it is likely that the spontaneous causes are mechanistic.

plained *despite* the fact that the world is rational, rather than *because* it is rational, for the claims of argument B are said to hold *even though* the world has received many blessings, including rationality, from the Demiurge (in BI, 269d8–9). It does not seem possible then that the participial expression ζῷον ὄν . . . (269d1–2) is meant to explain the reverse circuit of the universe or is to be read causally.[14] Rather the expression should merely be taken descriptively. The participial phrase is similar to expressions in the *Timaeus* (30b8, 36e4) and, like them, tells us only that the universe is endowed by the Demiurge with a rational World-Soul. No functional analysis of what role the World-Soul might play in the myth is stated or implied by the construction. Further, that the argument B proceeds *despite* the fact that the world is rational means that the purpose of the argument is not to explain the benefits which accrue to the world as the result of its being rational. The purpose is not to explain that the world's reverse motion is in fact initially uniform. Rather, the purpose of the argument is just to explain the cause of the reverse motion *qua* motion.

SECTION BI

If my analysis of the structure of the argument B is correct, the source of disorderly motion is to be found in BI. Here we are told that there is something which is most divine and that its attributes are denied to the phenomena just as being bodily.

There are two possible candidates for what is called "the most divine of all things": 1) the Platonic Ideas[15] and 2) the rationality of the Demiurge.[16] I will argue for several reasons that it is the Ideas that are intended.[17] The Ideas are called divine in both the *Phaedo* (81a5) and the *Republic* (500c9), and in the *Phaedrus* the Ideas are that in virtue of which the divine is divine (249c6). This derivative status of the gods to the Ideas in the *Phaedrus* would explain the superlative θειοτάτοις here in the *Statesman.*

Now phrases similar to those describing the properties attributed to the most divine here, namely being τὸ κατὰ ταὐτὰ καὶ ὡσαύτως ἔχειν ἀεὶ καὶ ταὐτὸν εἶναι, are elsewhere in the dialogues regularly used to

[14] So correctly Robinson, (1) 135n17. Herter takes the participial construction of 269d1–2 as his proof-text, and indeed only text, for the view that the cause of the reverse circuit of the universe is the rational World-Soul. Herter (2) 109.

[15] So Robinson (1) 132n6 and Guthrie 179.

[16] So Skemp (2) 105; cf. Skemp (1) 25n1.

[17] For two defenses of the traditional view that the theory of Ideas is present in the *Statesman,* see Guthrie 175–80, Mohr (1).

describe essentially *both* the eternal immutability of the Ideas (e.g., *Phaedo* 78c6, 80b2–3; *Republic* 479a2, *Sophist* 248a12; *Philebus* 59c4, 61e2–3; *Timaeus* 28a6–7, 38a3, 52a1) *and* the uniform motion which characterizes rationality (*Timaeus* 36c2, and especially, *Laws* X, 898a8–9).[18] So the phrase here (BI) is, just by itself, ambiguous. It should be noted, however, that whenever Plato uses similar phrases to refer to the Ideas, he does not use them attributively, i.e., as qualifying something further. The predicates rather are used to describe as best Plato can the *status* of the Ideas; they describe the formal, external, or metaphysical properties of the Ideas. When, however, he uses similar phrases in senses that would apply significantly to motions and not to the Ideas (as at *Timaeus* 36c2, 41d7, 82e1, where they mean "uniformly," "in the same mode," "in the same proportion"), it is always clear that the phrases are indeed being used attributively of motions, as indeed is the case in Ba where κατὰ ταὐτά qualifies κινεῖται. But this is not the case in BI. Plato could not have made it any clearer that he was describing a *state* of affairs rather than a type of motion than by using the neutral verbs he does use (ἔχειν . . . εἶναι). He could easily have clarified his meaning, however, if motion was intended, simply by replacing these verbs with κινεῖν or ἰέναι. Alternatively, if uniformity of motion was intended he might easily have clarified the matter by adding the qualification ἐν τῷ αὐτῷ as *Laws* 898a8 and as in Ba. For this qualification further describes uniformity of motion, but is entirely inappropriate to the Ideas, for Plato is explicit that in no sense does the relation ἐν hold of the Ideas (*Timaeus* 52a3, 52c6–d1).

Further, when the family of predicates describing the most divine is used of the circling of rationality in the *Laws,* the circling, as Lee has shown, functions as a kind of mass-term rather than a count-noun; the rotation is not the motion of circumnavigation but the rotation of the whole circle like the rotation of a wheel that spins in place.[19] Now this account of the predicates as applying to rationality makes them inapplicable to the Demiurge of the *Statesman* myth, since his rotations are explicitly viewed as countable (269c6) and repeated attention is drawn to the fact that his tasks are periodic and intermittent (269c–d, 272e, 273e).

Further and most decisively, when the related predicates assigned to the most divine are elsewhere in the dialogues denied of the phenomena, as they often are and as they are in BI, it is always by contrast to the Ideas that they are so denied and not by contrast to rationality. And when on the other hand the predicates are negated in the *Laws* passage (898b5–8), the negations hold not of the phenomena but of (irrational) souls.

---

[18] Robinson and Guthrie (above, n15) fail to take these passages into account.

[19] Lee (4) 76–78.

I suggest, then, that while the phrase τὸ κατὰ ταὐτὰ . . . ταὐτὸν εἶναι just by itself is ambiguous, it nevertheless refers here exclusively to the Ideas.

Now what the denial of these predicates to the phenomena means, however, is also ambiguous.[20] In some places in the dialogues the denial seems to mean only that an object may in one respect (in respect to a given time, location, observer, or partial aspect, cf. *Symposium* 211a) have one property and in another have an opposite or contrary property. This is clearly the case in "the argument from opposites" in *Republic* V (479a–c, cf. VII, 522c–525c) and in the closing argument for immortality in the *Phaedo* (102b–c, e; cf. *Hippias Major* 289a–c). In these cases the denial of immutability to the phenomena is merely the claim that phenomena are *capable* of motion, are *capable* of being qualified by opposites, are of no necessarily determinate quality. The phenomena, on this view, are disparate but not contradictory.

Elsewhere, however, the denial of immutability to the phenomena is clearly the much stronger assertion that the phenomena are in a universal, chaotic flux (e.g., *Philebus* 59b1, *Timaeus* 28a, 30a, 49a–c, 52e–53a, 58a–c). At *Phaedo* 90c4, for instance, when phenomena are said to be οὐδὲ βέβαιον, Plato means by this that "everything is driven and tossed this way and that, turned upside down, just as in a tidal channel where the flux and reflux is strong, and nothing ever remains in one place for any time" (90c4–6; see *LSJ*, s.v. εὔριπος).

The logical correlate to this state of affairs is to claim not just that the phenomena are capable of having contrary predicates in different respects, but that they are indeed subjects of self-contradictions. Twice in the *Phaedo* this much stronger claim is made that the phenomena, by sheer contrast to the self-identity of the Ideas (78e2–3), are never identical with themselves, are never self-consistent (αὐτὰ αὑτοῖς . . . οὐδέποτε, 78e3 [which is taken to be the fuller meaning of the denial of phenomena being κατὰ ταὐτα, 78e4]; μηδέποτε κατὰ ταὐτὰ . . . ἑαυτῷ, 80b4–5). The same logical claim is made of the phenomena in the *Timaeus*: all possible predicates, including contradictory ones, can be predicated of anything at any time (49b2–5). Now to deny the self-identity of the Ideas to the phenomena in this sense is not to claim that the phenomena in no way exist, but only to claim that they are in the process of change. In the *Parmenides* (155e–157b, and cf. 156e7–8 for the full generality of the doctrine), it is claimed that when a thing is changing from having a property to having its opposite, it passes through what is called "the suddenly" in

[20] For a different reading of what is entailed in the denial of these predicates, see Irwin, who claims the denials have a single sense and that Plato nowhere holds that the phenomena are in a universal flux. Irwin (2).

which it has neither the property nor its opposite. This is logically equivalent to a denial of the law of non-contradiction. And in the *Timaeus* Plato describes things in the process of changing as τὰ ἀνομοιούμενα ἑαυτοῖς (57c3–4). To say, then, that the phenomena lack the permanence and self-identity of the Ideas, in one sense, means to claim the phenomena are actually changing. And to claim that they are *never* αὐτὰ αὑτοῖς and *never* κατὰ ταὐτὰ ἑαυτῷ (cf. also *Philebus* 59b1) is to say they are in a universal flux.

Of the two options of interpreting the denial of permanence to the phenomena in the *Statesman,* clearly the first, namely the view that the phenomena differ in different respects and are merely *capable* of change not only is simply irrelevant to the argument, which is trying to establish a cause for the reverse circuit, but also ensures only the possibility of motion without guaranteeing the actuality of any motion. Since the further claim is immediately made (in BI) that because the world is bodily, it cannot be free of change, what is required is that the denial of permanence to the phenomena entail actual, not just potential change. So the first alternative interpretation of the denial of permanence to the phenomena fails.

It seems that the second sense, namely, that the phenomena are contradictory and in flux, is the sense of the denial that is intended and that fulfills the demands of the context. The claim of BI, then, seems to be that the phenomena just on their own are the cause of the reverse circuit, or of what it stands for on a non-literal reading—disorderly motion. This is reinforced by the addendum that the Demiurge and his minions, the World-Soul, other souls, and their rationality, are all irrelevant to explaining the cause of the reverse circuit (BI, 269d8–9). Moreover, it seems the Demiurge and his minions are not even primarily being viewed as efficient causes of motion; their function is not to stir up motion, it would seem, but rather to retard and stabilize the phenomena, since the motions of the corporeal are said to exist *in spite of* these various agents (ἀτὰρ οὖν δή, 269d9). I suggest, then, that when in A it is said that the reverse circuit is αὐτόματον (269c7), this means that the phenomena move on their own without any assistance from external agents.

The particular contribution of this *Statesman* passage to other passages in the dialogues which assert that the phenomena, in contrast to the Ideas, are in flux, is that while the others assert that the phenomena *necessarily* are in flux, the *Statesman* in addition asserts that they *on their own* are the cause of their being in flux. The *Statesman* asserts that they are necessarily and *essentially* in flux. Now the *Statesman* does not go on to give a mechanism by which it is explained how the phenomena do maintain on their own a chaotic flux; this task, though, is accomplished in the *Timaeus* (58a–c, on which see chapter 6).

SECTION Ba

The γε μήν of 269e2 is strongly adversative. Having established the counter-course of the universe *qua* motion, Plato goes on to suggest that, despite our expectation that in the absence of the Demiurge the motion would be disorderly (cf. *Timaeus* 53b), the counter-motion will in fact to a large, but qualified (κατὰ δύναμίν), extent be the uniform motion of rotation in a single place (direction?). I take the shift from τὸ ἀνάπαλιν ἰέναι (A, 269d2) to τὴν ἀνακύκλησιν (Ba, 269e3) as significant. The emphasis here (Ba) is on the circularity and uniformity of the counter-motion, but neither the cause of uniformity nor the reason for its partial failure is here explained. These explanations are only forthcoming at 272e–273e and are so delayed, I suggest, because they are in fact irrelevant to the argument which B is trying to establish, namely, the efficient cause of disorderly motion.

That Plato here (Ba) and later (272e–273e) draws special attention to the uniformity and circularity of the reverse course causes trouble for those critics, like Skemp, who claim that the efficient cause of the reverse circuit is an irrational World-Soul or an irrational element in the World-Soul which, as irrational, could only cause disorderly motion. The best Skemp can do is attempt to explain away the regularity of the reverse circuit as an exigency of myth's structure.[21] This attempt is unconvincing when compared to explanations which "save" rather than destroy the appearance of the initial orderliness of the reverse circuit (see below).

Unlike most critics, I take εἴληχεν at 269e4 in an impersonal sense to mean "obtain by lot or by fate" rather than "obtain by the will of the gods" (see *LSJ*, s.v. λαγχάνω), and not as referring to the Demiurge. In the *Statesman* myth there is a sense of an impersonal, non-temporal, fate or destiny behind every component of its ontological scheme.[22] The Demiurge, admittedly, is himself a personal source of destiny for some things. He is explicitly the source of the destined aspect of the World-Soul (φρόνησιν εἰληχὸς ἐκ τοῦ συναρμόσαντος αὐτό, 269d1). However, the Demiurge himself seems to be subject to certain destined restrictions not of his own making (τοῦ προσήκοντος αὐτῷ μέτρον . . . χρόνου, 269c6–7). This thought is even more clearly enunciated when it is said that it is not θέμις (269e7) for the Demiurge to have two contrary motions.

There is nothing, therefore, in Ba or elsewhere in the argument to suggest that the reverse circuit is either directly or indirectly caused by the Demiurge. Even if εἴληχεν did, however, refer to the Demiurge, it would

---

[21] Skemp (1) 27, (2) 89.

[22] Cf. Cornford (2) 208.

not entail that he was the cause of the motion of the reverse circuit but only entail that he was the cause of its regularity.

SECTION Bb

Section Bb adds information to that already given in A which will be used to eliminate the later disjuncts (BII, III, IV). Two points are made. First, we are told that the Demiurge, who guides all things that move αὖ, is the only thing which always turns itself (Bbi). This has been taken to mean that the Demiurge is an autokinetic soul and that the motion of the phenomena (and indeed the counter-motion of the universe) has its source ultimately in the Demiurge.[23] However, the αὖ here, I suggest, means only "backwards" rather than "in turn." For by this point we have repeatedly been told that the universe does have a reverse course and this expression is the briefest way of describing that. So the phrase τῶν κινουμένων αὖ πάντων (269e6) does not mean that the motion of the phenomena is being viewed as essentially transitive and not as essentially spontaneous.[24] Rather, I suggest, the eternality and reflexivity of the Demiurge's turning mark him as the only entity whose rotation is independent and non-contingent. Any other rotation, whether of the phenomena or of the World-Soul or souls generally, is derived and contingent. I take it that the reflexivity and eternality here do not simply mean that the Demiurge is the only thing that can circle in the same *direction*, for this is the second point made in Bb and this second point is not viewed as explaining the first (δέ, 269e6).

The second point made in Bb is that it is contrary to divine law (οὐ θέμις) for the Demiurge to move at different times in different directions (Bbii). A similar appeal to impersonal divine law is made in the *Timaeus* where it is said that it is never θέμις for the Demiurge to do anything other than the best (30a6–7). The reason in the *Statesman* that the Demiurge cannot have contrary motions is two-fold. First, the motions which the Demiurge would set up by moving in opposite directions would be erratic and he is only a source of order and goodness. Second, to so move would draw into doubt his essential and eternal rationality, which is characterized by uniform motion in a single direction.

The image of the Demiurge that emerges here (and from 272e–273e, to glance ahead) is that the Demiurge operates somewhat like a mechanical clutch. When he is in contact with or engages the world, he imparts both his rotation, i.e., his motion as circular, and the direction of his motion to

---

[23] Robinson (1) 133, 135, Cherniss (6) nn21, 44.

[24] Against αὐτὸ ἑαυτὸ στρέφειν ἀεί referring to autokinesis, see above, n3.

the world; but he can disengage from the world and still continue with his motion and direction, while the world, then, goes on its own way.

SECTIONS BII, III, IV

With Bb in tow, Plato is ready to state and eliminate what he takes to be an exhaustive set of remaining possible explanations of the reverse rotation of the universe.

The first possibility which is entertained and dismissed is that the orderly world always turns itself (BII, 269e8–9). Now we can presuppose as already established that there is a reverse circuit of the world (269e3). So when it is denied that the cosmos is always self-turning, we must suppose that the direction of the hypothesized turning is in the direction of the retrograde cycle of the universe. The denial, then, does not seem to be that the world is in no sense self-moving, but is only that the cosmos cannot maintain on its own orderly motion. The *kósmos* here (269e8) seems to be the ordered World-Body without necessarily including the World-Soul, for at 269d8–e1 (in BI) *kósmos* is contrasted both to the corporeal as such and to the World-Soul which it receives from the Demiurge. The schema that is being denied here (BII, 269e8–9), then, is that of the Demiurge eternally revolving in one direction outside of the universe and the universe itself spinning eternally in the opposite direction, both circuits being completely independent of each other. The passage adumbrates 273a ff. where the cosmos when independent of the Demiurge begins to decay, because the cosmos cannot eternally remain uniformly in motion. Only by the Demiurge's intervention can it be restored to order. The denial of the world's eternal self-turning is a direct corollary of the fact that eternal self-turning is an exclusive possession of the Demiurge (Bbi, 269e5–6).

The next possible explanation of the reverse circuit is that the Demiurge always moves the world as a whole in two directions (BIII, 269e9–270a1). The denial is obviously meant to be a direct corollary of Bbii, namely, that the Demiurge cannot move now in one direction, now in another. It would seem, therefore, that Plato is not even entertaining the possibility (let alone affirming it) that anything but immediate direct effects are going to count as possible causes of the reverse circuit; this disjunct shows that indirect causes, such as those Cherniss wishes to find as the cause, are not even considered.

The final possible explanation of the reverse circuit is that there are two gods who turn the world with conflicting purposes (BIV, 270a1–2). The denial of this view is a denial of Zoroastrianism. Zoroaster is mentioned by name in *Alcibiades* 1 (122a1). The reason for the denial of this view is the same rationale that stands behind Bbi, namely, since for Plato it is inconceivable that gods are sources of evil, it is therefore impossible for them to have conflicting purposes.

With these competing disjuncts dismissed, the argument returns to and asserts by the force of the disjunctive syllogism the earlier alternative BI, that the corporeal is the efficient cause of the reverse circuit (270a2–8). The description given in this return must be read in light of the alternatives already given and does not itself state for the first time the cause of the reverse circuit. The cause is merely summarized. This occurs in the line "but at another time when it [the world] is let go, it moves itself throughout itself" (C, 270a5–6). This looks directly forward to the description of the chaos that ensues upon the Demiurge's withdrawal from the world, described at 273a: "And as the universe was turned back and there came the shock of collision, as the beginning and end rushed in opposite directions, it produced a great earthquake within itself" (Fowler).

The new information conveyed in section C is that the reverse circuit lasts for a very long time. This is explained (διά, 270a7) by means of a mechanical model. It should be noted *(contra* Cherniss) that this model is not given to explain the efficient causality of the reverse circuit; that, at this point, is assumed. Those, like Skemp, who seek a non-mechanistic explanation of the reverse circuit similarly err. For they feel the need to *explain away* the model as "an appeal to the example of a familiar piece of apparatus [which] may help to justify a piece of cosmology which was bound to strike astronomically-minded hearers as—unlikely, to say the least."[25] The apparatus entertained by Skemp and others is a globe suspended from a twisted string, which as it unwinds suggests the retrograde cycle of the universe. Cherniss takes the passage so interpreted as his proof-text for the view that disorderly physical motions (represented by the reverse circuit and reflected in the model as the alleged unwinding of the thread) are secondary spin-off effects of the Demiurge's primary organizational efforts (reflected in the alleged winding up of the thread).[26] I suggest, though, that this is not even the model Plato has in mind. The mention of the small pivot (270a8) and the lack of any mention of a string or wire strongly suggest that Plato is thinking here of a spinning top (as at *Republic* 436d) rather than a suspended globe. Further, the model of the suspended sphere does not fit or explain the purpose or details of the passage. The suspended sphere is taken as an image for how the world might move on its own. But this is not what is required of the model. What is required is an explanation of why the world revolves for such a long period on its own. Further, the details we

---

[25] Skemp (2) 102, cited with approval by Robinson (1) 137n23, and by Herter (2) 109n13.

[26] Cherniss (6) n21.

are given of the mechanism, namely, that it is very large, evenly balanced, and on a very small pivot, are all irrelevant to the image of an object suspended from a twisted thread. Whereas all these details are exactly relevant in explaining why a top would spin many times on its own before giving over to erratic motions.[27]

With this analysis of the argumentative section completed, the rest of the myth falls fairly easily into place. One troublesome point of interpretation is determining the reference of the expression σύμφυτος ἐπιθυμία (272e6). This εἱμαρμένη τε καὶ σύμφυτος ἐπιθυμία is said to turn the cosmos back again when the Demiurge withdraws from the world (272e5–6). At first blush the expression seems strong literal evidence for those who wish to see an irrational psychic force as the cause of the reverse motion of the universe.[28] It is the mainspring of Plutarch's reading of Plato's cosmology. My sympathies, though, lie with Vlastos, who sees this as a colorful expression for a purely physical impetus.[29] Herter is correct to see the expression as having a backward reference to a similar expression in an identical context: ἐξ ἀνάγκης ἔμφυτον (269d2–3, in A).[30] The two expressions seem merely stylistic variants of each other. Herter, however, mistakenly interprets the earlier expression by reference to the allusion to the World-Soul at 269d1 (in A), a line which he, as we have shown, mistakenly reads causally rather than descriptively. Rather, the phrase ἐξ ἀνάγκης ἔμφυτον is to be explained in the argumentative section B; but in B there is no mention of an irrational soul, or irrational aspect of a soul, as the cause of the reverse circuit. Further, the term ἐπιθυμία is used loosely in a non-technical sense elsewhere in the myth (272d3). We can, then, with some confidence say that ἐπιθυμία (272e6) is a poetic expression for what is indeed not a psychic force. Moreover, the

---

[27] Brumbaugh has tentatively suggested that the mechanical model described at 270a is a precursor of weight-and-water operated clocks and that this accounts for the model's possessing a foot or pivot, 523–24, 525–26. This view has in turn tentatively been accepted by Skemp (1) 129–130. This model, however, fails to explain why the pivot's being small and why the pivoted body's being large and evenly balanced account for the long duration of the reverse circuit, since on the model of the weight-and-water operated clock the length of duration of the clock's running on its own is dependent solely on the size of the clock's well-shaft and the rate at which water runs out of the shaft. See Brumbaugh 526n30. Like those critics who take the model to be a suspended globe, Brumbaugh mistakenly assumes the mechanical model is invoked to help *explain* the cause of the reverse circuit.

[28] Skemp (1) 26-27, cf. (2) 107–108; also see Robinson's qualified acceptance of Skemp's interpretation, (1) 136–38 and 135n18; and see Herter (2) 111–12.

[29] Vlastos (2) 395.

[30] Herter (2) 111.

phrases ἔμφυτον and ἐπιθυμία seem to adumbrate and form a group with the expression "that which is inherent in the primeval (i.e., acosmic) nature" (τὸ . . . σύντροφον) at 273b4–5, which is set in apposition to "the corporeal" (τὸ σωματοειδές, 273b4) and explains the cause (αἴτιον) of the World-Soul being disrupted.

I agree with Herter that matter, which is itself moving chaotically, infects and exerts pressure on the World-Soul.[31] Robinson, on the other hand, claims that though the bodily is that which accounts for the evil in the world, the bodily itself is not as such positively evil. He compares the bodily to a virus, which in itself is inert and neutral, and which is active and harmful only when in contact with a living organism.[32] This image, however, does little justice to the quite active, even explosive disruptions by which the corporeal, when the gods withdraw from the world, temporarily but completely throws out of kilter the organizing ability of the World-Soul (273a–b). Robinson correctly sees that in this passage Plato is applying on a cosmic scale the same corruption of the soul by the corporeal as is described in the discussion of the newly incarnate infant soul in the *Timaeus* (43a ff.).[33] The parallelism is striking. It extends to such details as the bodily in the discussion of the infant soul, as in the *Statesman* myth, being said to flow in the opposite direction to the circuits of rationality (*Timaeus* 43d2–3). But *contra* Robinson, in both the *Timaeus* and the *Statesman* myth the incursion of the bodily is hardly that of a neutral, inert, and impassive substance. In the *Timaeus* the soul is confined within "the flowing and ebbing tide of the body" (43a5–6), which is described as a "strong river" (43a6). The result of the confinement of the soul in the body is that the soul suffers violent motions (43a7, cf. 43c7–d2) and loses its ability to maintain order (43b1, cf. 44a4). Further, sensations, which are treated as incursions of the corporeal into the soul, are said to assail the soul and cause a yet greater tumult (43b6–c7). The parallelism of *Statesman* 272e–273e to the infant psychology of the *Timaeus* strongly suggests that the bodily condition which infects the World-Soul is not inert and passive, but is actively chaotic.[34]

The World-Soul of the *Statesman*, because it remembers the teachings of the Demiurge, is able to restore order to the world after the initial active incursion of the bodily into the World-Soul (273a7) and so leads it from a state of σεισμός (273a3) to a motion of *uniform* rotation (εἰς δρόμον κατακοσμούμενος, 273a6–7). The universe, however, despite its shaking, is already said to be on its reverse circuit (ὁ μεταστρεφόμενος)

---

[31] Herter (2) 111.
[32] Robinson (1) 136-37.
[33] Robinson (1) 138.
[34] Cf. Vlastos (2) 397.

when the World-Soul asserts its influences (273a1–2, 4), so again it does not seem possible that the World-Soul is being viewed as the cause of the reverse circuit.

The organization which the World-Soul is able to restore to the universe is temporary. For the bodily again begins to assert itself, with the result that, as in the infant psychology of the *Timaeus,* the soul's rational faculties begin to break down. In the *Statesman* this is characterized by loss of memory (273b3, c6). The primeval bodily condition (273b5, b7–c1, c7–d1), however, is the cause (αἴτιον) not only of the World-Soul's forgetfulness but also of all difficulty and injustice which arises in the universe (273c1). Now it is true that disorder (ἀταξίας, 273b6) and lack of harmony (ἀναρμοστίας, 273c7), which characterize the ancient bodily condition, could qualify either static or dynamic states. But insofar as the ancient condition, when it holds power (δυναστεύει, 273c7), is said to burst forth (ἐξανθεῖ, 273d1) resulting in the world's being storm-driven (χειμασθείς, 273d5) at sea, it again is fairly clear that the disorder of the ancient state is a dynamic one.

With the forgetfulness of the World-Soul comes an increase in the disorder of the corporeal. There is no suggestion, however, that the World-Soul contributes to this increased disorder as a proximate efficient cause. In this the *Statesman* myth differs from the infant psychology of the *Timaeus,* where the once rational infant soul, when made irrational by the incursions of the bodily, in turn causes violent motions in the corporeal (43a7–b2). Rather, in the *Statesman* myth the World-Soul seems simply to relinquish its care and rule over the universe as a result of its forgetfulness, such that the corporeal may on its own, in the absence of organizational restraints from the World-Soul, revert to its ancient state of chaos. It seems, then, that as the World-Soul's memory is erased, the World-Soul does not become actively irrational, but simply becomes non-rational. With the increasing forgetfulness of the World-Soul, its activities do not become erratically exuberant but become increasingly dull and lacking in energy (ἀμβλύτερον, 273b3). In this case, again, the World-Soul is not being viewed primarily, if at all, as an efficient cause of motion but is being viewed mainly as a maintainer of organization (which it transmits to the World-Body from the Demiurge) against a natural tendency of the corporeal to be chaotic (see chapter 9). But, the World-Soul here is capable of maintaining order only within a certain range of natural disruptions. Chaos wins out.

The primitive condition of the corporeal, towards which the dissolving (διαλυθείς, 273d6) universe tends, is characterized by unlimited dissimilarity (τὸν τῆς ἀνομοιότητος ἄπειρον, 273d6). This characterization of the primeval corporeal condition looks very much like the characterization of the corporeal as such in BI (269d5–7), where self-identity is denied of the corporeal.[35] Since no psychic cause is found in 272e–273e for the reverse circuit of the universe and since, as we have seen, the primitive condition of the corporeal is a dynamic state, the two characterizations of the corporeal as being denied self-identity mutually support the view that the corporeal in and of itself is in chaotic motion.

The cosmological schema which emerges from our analysis of the *Statesman* myth, then, is that the corporeal on its own without psychic influences moves chaotically, is the efficient cause of the retrograded cycle of the universe, and is a positive source of disorder and evil. That the retrograde motions caused by the corporeal is uniform, at least initially, is the result of the ability of the World-Soul to sustain organization. That the World-Soul has such an ability should dissipate the concern of some critics that the σύμφυτον ἐπιθυμία, whether interpreted as an erratic physical or irrational psychic force, could not be the source of motion which is uniform.[36] The World-Soul itself, however, seems not primarily to be viewed as an efficient cause either of motion as such or of motion as regular or erratic. The World-Soul is simply a *maintainer* of the organization of both itself and the World-Body (273a7–b1). The Demiurge, which stands over and above the World-Soul and which is the source of its existence, is the *source* of the orderliness of both the World-Body and World-Soul, which the World-Soul tries to maintain against the inherently chaotic motions of the corporeal.

---

35 This characterization also looks very much like an adumbration or echoing of the *Timaeus,* where the chaotic flux of the phenomena is described by saying the pre-cosmic receptacle is full of entirely dissimilar powers (μήθ' ὁμοίων δυνάμεων, 52e2). I think the reading of the *Statesman* mss τόπον (273e1) should be retained over Proclus' extension of the passage's marine metaphor, πόντον. The phrase "the *place* of unlimited dissimilarity" may in fact be a blind reference to the receptacle (cf. ἔν τινι τόπῳ καὶ [= i.e.] κατέχον χώραν τινά, *Timaeus* 52b4–5).

36 Herter (2) 111, Robinson (1) 137.

EIGHT

The Sources of Evil Problem and the Principle of Motion Doctrine in the *Phaedrus* and *Laws* X*

In 1939 Vlastos suggested that the descriptions in the *Statesman* and *Timaeus* of disorderly motions which apparently have no psychic causes could be reconciled with the doctrine from the *Phaedrus* (245c9) and *Laws* X (896b1) that soul is the source and cause of all motions, by limiting the scope of the ἀρχὴ κινήσεως doctrine. Soul is the source of all motions, Vlastos claimed, only in the demiurgically ordered world; the doctrine's scope does not cover the acosmic periods of the *Statesman* myth (273a–e) and the *Timaeus* (30a, 52d–53a): "The proposition that the soul is πρῶτον γενέσεως καὶ φθορᾶς αἴτιον (*Laws* 891e) merely denotes the supremacy of the soul's teleological action *within the created universe*."[1]

Five years later Cherniss found this view so incredible as to be refuted simply by having attention drawn to it: "Vlastos, seeking agreement between the *Laws* and a literal interpretation of the *Timaeus*, has to insist . . . that where in the former Plato calls soul the principle of all motion he is not to be understood as meaning literally 'all' (!)"[2] And, indeed, Vlastos quietly dropped this view by 1964, at which time he abandoned any attempt to reconcile the *Timaeus* with the ἀρχὴ κινήσεως doctrine,[3] but not without first having initiated a tradition of critics who have accepted his earlier stance.

Under the direct influence of Vlastos' 1939 article, Hackforth in a posthumous work claims that "as to the ungenerated soul of the *Phaedrus*, I do not believe that this should be taken as inconsistent with a pre-cosmic absence of soul . . . . We may fairly take him [Plato] to mean that soul, being the necessary presupposition of all movements that occur in the universe [i.e., the *formed* universe], is co-eval with the universe."[4]

---

* Reprinted from *Apeiron* 14 (1980) by permission of the editors.

[1] Vlastos (2) 397, his emphasis.

[2] Cherniss (1) n362.

[3] Vlastos (3) 414–15, cf. (2) 396n4.

[4] Hackforth (3) 21.

Vlastos' limitation of the ἀρχὴ κινήσεως doctrine to the formed world has also been accepted and argued for by Robinson[5] and Easterling.[6,7] Though I agree with Cherniss that this view has a *prima facie* implausibility, its rising popularity suggests that the arguments advanced in its favor deserve an examination which they have not received to date.[8] I will argue that the arguments for the limitation of the doctrine are faulty.

The arguments of Vlastos, Hackforth, and Robinson for the limitation of the scope of the ἀρχὴ κινήσεως doctrine are largely arguments from the economy of Plato's literary-argumentative style. They all wish to claim that Plato does not explicitly make the pre-cosmic chaos an exception to the scope of the ἀρχὴ κινήσεως doctrine because he does not need to make mention of the pre-cosmos in dealing with the specific problems which he uses the ἀρχὴ κινήσεως to help solve. Thus in speaking of the *Laws* X, Vlastos writes: "That primary causation might rest with the evil soul . . . is forthwith declared to be contrary to fact, and the speaker can go on to complete his case against the atheists, without digressing to explain how primary causation through the evil soul is, in fact, inexplicable save through collision with material, secondary causes [i.e., non-psychically caused erratic motions which impinge upon the soul as in the case of the infant soul (*Timaeus* 43a–b) and the World-Soul (*Statesman* 273b)]."[9] Thus too Hackforth claims of the argument for immortality in the *Phaedrus*, as his sole justification for limiting the scope of the ἀρχὴ κινήσεως doctrine, that "'chaos' is outside Plato's purview in considering the nature of ψυχή here."[10] And thus too Robinson writes: "In the *Phaedrus* and *Laws* the argumentation is very generalized and schematic: no

[5] Robinson (1) 95–97, (2) 348–50.

[6] Easterling, especially 31, 37–38.

[7] Unless one wishes to resort to Plutarch's belief in a pre-cosmic evil World-Soul, the limitation of the scope of the ἀρχὴ κινήσεως doctrine to the ordered world is the only way that a literalist reading of the creation myth in the *Timaeus* could be reconciled with the ἀρχὴ κινήσεως doctrine. It is no accident that Vlastos, Hackforth, Robinson, and Easterling all read the creation myth literally. This mode of reconciliation would also apply to a non-literalist reading which, correctly in my view, did not find a psychic cause for the erratic chaotic factor which the pre-cosmos represents on a non-literalist reading.

[8] Tarán has made some incisive criticisms of Easterling's arguments. Tarán n131.

[9] Vlastos (2) 397. Vlastos here is implicitly giving an interpretation to the expression ἀνοίᾳ συγγενομένη (*Laws* 897b3) which clearly goes beyond the text. The point of 897b is that soul in and of itself is ethically neutral; see Robinson (1) 149. How soul comes to take on an ethical cast is left completely open in this passage.

[10] Hackforth (3) 21.

reference is made to pre-cosmic states, and understandably so, since the context does not demand them."[11]

Though arguments from economy of literary style are standard equipment for most Platonic scholars and are often warranted given Plato's dialectical concerns, they are the most difficult of arguments both to establish and to refute. Even Robinson is willing to admit "that the *argumentum e silentio* is often a foundation on quicksand."[12] Nevertheless, we are not completely at a loss in evaluating such arguments. I think a minimum requirement for the success of an argument from silence is that the doctrine which is claimed by such an argument to be implicit in a text, but omitted there for reasons of literary or argumentative economy, must be found explicitly stated somewhere in the Platonic corpus. And yet, Vlastos', Hackforth's, and Robinson's arguments from silence fail this simple criterion. Nowhere does Plato explicitly state that the ἀρχὴ κινήσεως doctrine applies only to the ordered world. And yet one would expect that if Plato had meant to limit the scope of the doctrine, he would have made some mention of it. For consider the following.

It is clear from the incursions of the bodily into the World-Soul in the *Statesman* myth and from the description of the infant soul in the *Timaeus*, and more generally from the never surmounted limitations of the Demiurge (especially *Timaeus* 30a3) that the chaotic flux of the phenomena in the acosmic periods of the *Statesman* and *Timaeus* continues to be a factor within the ordered universe. Therefore, critics who claim the chaotic flux of phenomena, which constitutes the pre-cosmos, has no psychic cause would also have to claim that even within the ordered world the flux of phenomena would have to be exempted from the scope of the ἀρχὴ κινήσεως doctrine.[13,14] Given this further need to explain the

---

[11] Robinson (1) 95, (2) 348.

[12] Robinson (1) 95. Robinson wishes to claim that, nevertheless, "in this particular case it [the argument from silence] seems to have more cogency than usual." Robinson (1) 95. In trying to establish this cogency, Robinson, successfully in my view, argues against Plutarch's suggestions that a pre-cosmic irrational psychic force causes the chaotic motions of the pre-cosmos. Robinson (1) 96–97, (2) 348–49. This, however, is not sufficient to establish by disjunction the limitation of the ἀρχὴ κινήσεως doctrine. On Robinson's other arguments for restricting the scope of the ἀρχὴ κινήσεως doctrine, see n14 below.

[13] So correctly observes Herter (1) 343.

[14] Robinson, perhaps anticipating this bind, emphasizes the differences between the pre-cosmos and the ordered world in the *Timaeus*, so that he can claim that "Plato . . . is dealing in two instances with two completely different types of motion, the one accepted and universally admitted, and operating in an organized world of temporal succession, the other a reality in every sense different, and almost beyond description." Robinson (1) 97, (2) 349–50. In support of this

limitation of the scope of the ἀρχὴ κινήσεως doctrine at least in the *Laws*, written after the *Timaeus* and *Statesman*, the argument from silence is considerably weakened.

But even aside from this methodological failing of the argumentation for the limitation of the ἀρχὴ κινήσεως doctrine to the ordered world, there are two sorts of evidence which suggest that Plato could not have intended such a limitation of the doctrine: I) among the motions which Plato claims to be caused by soul in the *Phaedrus* and *Laws* are included the very motions by which he characterizes the pre-cosmos in the *Timaeus*, and II) the existence of non-psychically caused motions would destroy the force of both the argument for immortality in the *Phaedrus* and the argument for the existence of autokinetic soul in the *Laws* X. I will now try to establish I) and II) for the *Phaedrus* and *Laws* X.

BODILY MOTION AND IMMORTALITY IN THE *PHAEDRUS* (245c–246a)

Recent critics have, by and large, failed to understand the structure of the argument for immortality in the *Phaedrus*. There is a growing tendency among critics to take τὸ γὰρ ἀεικίνητον ἀθάνατον (245c5) not as

---

claim that there are "two completely different types of motion," Robinson argues 1) that in the absence of time, which the Demiurge is said to create, temporal succession is not possible and so *a fortiori* that motion as we know it is not possible and 2) that "the bodily" is not present in the pre-cosmos and so *a fortiori* motion as we know it is not present in the pre-cosmos. Against 1), see chapter 2, section II. The Demiurge does not create temporal succession, rather he creates the measures of temporal succession—clocks.

Concerning 2), the contents of the receptacle are explicitly called "bodies" (σώματα, 50b6) in the pre-cosmos; the reference to the pre-cosmos at 48b3–5 governs the whole discussion up to the first demiurgic interventions into chaos at 53b. Cornford, whom Robinson follows in this matter—Robinson (1) 94—tries to explain away this occurrence of σώματα as describing the contents of the receptacle in the pre-cosmos simply on the weight of his interpretation of 49b–50a. Cornford (2) 181. For an interpretation of 49b–50a which does not exclude bodies from the receptacle in the pre-cosmos, see chapters 3 and 4 above. Further, from 58a–c and its backward reference to the pre-cosmic chaos at 52d–53a, it is clear that the motions of the pre-cosmos are simply the rectilinear motions of the four primary bodies, motions which persist even in the formed cosmos. But in any case, the ἀρχὴ κινήσεως doctrine is explicitly said to cover conditions prior even to the length, breadth, and depth of bodies (*Laws* 896d; cf. 894a).

Plato is not then "dealing in two instances with two completely different types of motion" in the *Timaeus*, one covered by the ἀρχὴ κινήσεως doctrine and one not.

a premise but as a conclusion in the argument's structure.[15] This is shown to be wrong, though, by the ensuing period: τὸ δ' ἄλλο . . . ζωῆς (c5–7), which goes uninterpreted by Verdenius and Ackrill and is under-interpreted by Robinson, who takes it to mean "if it [that which is always moving] were not self-moving, there would be a cessation of life and movement."[16] To take the sentence in this manner, making it exactly equivalent in sense to the ensuing sentence (μόνον δὴ . . . κινούμενον, c7–8) is to dissolve any contribution which the phrasing τὸ ἄλλο κινοῦν καὶ ὑπ' ἄλλου κινούμενον (c5–6) (hereafter denoted as "A") might add to the argument and to miss the contrast of this phrase to the phrase τὸ αὑτὸ κινοῦν (c7) (hereafter denoted as "B"); the former ("A") is not simply the logical negation of the latter ("B"), as Robinson implies. Rather A (that which causes motion [in another] and is [itself] moved by [yet] another) is advanced as a possible candidate for being τὸ ἀεικίνητον (c5). It fails at this because it is capable of ceasing to move. This failed candidacy, though, suggests: 1) that the manuscripts' ἀεικίνητον is to be retained over the αὐτοκίνητον of the Oxyrhynchus Papyrus 1017 (which was adopted by Robin and Ackrill, et al.)[17] 2) that δέ (c6) is adversative[18] and 3) that τὸ ἀεικίνητον ἀθάνατον is being treated axiomatically[19] and is being taken in the argument as a criterion: we are looking in the argument for something which is ἀεικίνητον, and so therefore is also immortal. The link between eternal motion and soul will be made, as the expression παῦλαν ἔχει ζωῆς (c7) hints, by way of the concepts of life and animation.

The formulation "A" helps establish point I) for the *Phaedrus*, namely, that among the motions which Plato claims to be caused by soul in the *Phaedrus* and *Laws* are included the very motions by which he characterizes the pre-cosmos in the *Timaeus*. Given that B is eventually cashed out as the ἀρχὴ κινήσεως, A must ultimately be moved by B. Yet ἄλλου (c6) need not denote B, since objects characterized by A are themselves said to cause motion (κινοῦν, c6) (in other things also characterized by the description "A"). "A," then, merely describes transitivity of motion. The contrast between A and B is re-introduced later in the argument when that which moves itself (B) is identified with soul (245e4–246a1). Here the contrast is between ἔξωθεν τὸ κινεῖσθαι (e5) (= A), which characterizes the bodily as such, i.e., as soulless (πᾶν σῶμα ἄψυχον) and ἔνδοθεν

---

[15] So Ackrill 278, Verdenius 276, and Robinson (1) 112n7.

[16] Robinson (1) 112.

[17] Cf. Robinson (1) 111. As Skemp has pointed out, the reading αὐτοκίνητον would make c6–7 otiose. Skemp (1) 3n2.

[18] Against Verdenius, 276, and Robinson (1) 112n7. The particle thus emphasizes the failure of the subject of the sentence to be that which is always moving.

[19] On its axiomatic nature, see Skemp (1) 3n2.

αὑτῷ ἐξ αὑτοῦ <τὸ κινεῖσθαι> (e5–6), which characterizes body as animated by soul.

Now no one would deny that this transitivity of motion which characterizes the bodily as such does not fall under the scope of the ἀρχὴ κινήσεως doctrine. And yet this very transitivity of motion is introduced in the discussion of accompanying causes in the *Timaeus* (46c–e) in much the same wording and with much the same purpose, namely, to characterize the bodily as such.

In the *Timaeus*, the backward reference from 47e3 to the accompanying causes and the description of the accompanying causes which "produce their sundry effects at random and without order" (46e5–6, Cornford's translation) leave no doubt that these are the causes which operate in the pre-cosmos (cf. 30a4–5). These accompanying causes are contrasted with (rational) psychic causes and are the causes which belong to the primary bodies (earth, air, etc., 46d2–3 with d5 and e1–2).[20] The accompanying causes "arise by being moved by other things and in turn of necessity move still other things" (ὅσαι ὑπ' ἄλλων μὲν κινουμένων, ἕτερα δὲ ἐξ ἀνάγκης κινούντων γίγνονται, 46e1–2).[21] The transitivity of motion here used to characterize the bodily as such in the pre-cosmos of the *Timaeus* is the same as that used in the *Phaedrus* to characterize the bodily as such. It seems unlikely then that the ἀρχὴ κινήσεως doctrine of the *Phaedrus* is not meant to cover the types of motion which characterize pre-cosmic chaos. Our point I), then, is established in the case of the *Phaedrus*.

Let us return to the argument for immortality to show that point II) also holds of the argument, namely, that if there were pre-cosmic motions which were not initiated by soul, the argument would collapse.

With the failure of A to fulfill the criterion of eternal movement, another candidate is entertained, namely, B (245c7–8; δή emphatic but adversative, contrasting A and B). B is said never to cease from moving ἅτε οὐκ ἀπολεῖπον ἑαυτό (245c7–8). I do not take this sentence as in itself ending one argument for immortality and 245c8 ff. as starting a new independent apodictic argument (as does Robinson).[22] Rather, I suggest that what follows (245c8 ff.) is an elaboration of what it means for that which moves itself not to break contact with itself, such that it is immortal. This is done by showing that B neither comes into existence nor perishes, *given* that it does not break contact with itself. The expression οὐκ ἀπολεῖπον ἑαυτό (c8) does not seem to mean "it cannot abandon its own

[20] The participles of *Timaeus* 46d2–3 are causal and quasi-instrumental in force (cf. e6); Cornford correctly translates as "producing effects by."

[21] Taking κινουμένων(e1) subjectively (Cornford) or objectively (Skemp) with ὅσαι <αἰτίαι> makes γίγνονται merely equivalent to εἰσί and would also seem to require a definite article. I take κινουμένων as part of a genitive absolute.

[22] Robinson (1) 113.

nature," for in that case it, as Hackforth points out, would apply equally well to A.[23] Rather it seems to mean that B indeed does not cease moving itself, does not cease causing its own motion.[24,25] Now this condition of B, that it does not cease moving itself, goes onto the back burner until it is needed at d5–6.

In the meantime, the argument proceeds from the claim that B is also the source and principle of motion for whatever else moves (c8–9). This claim is stated as an axiom.[26] With the axiom that B is an ἀρχὴ κινήσεως and the axiom that everything that comes to be comes to be from an ἀρχή (d1–2, 6; cf. *Timaeus* 28a, *Philebus* 26e) it is shown that B is ἀγένητον.[27,28]

The proof that the ἀρχὴ κινήσεως is imperishable is contained in the following period:

i) ἀρχῆς ἀπολομένης
iia) οὔτε αὐτή ποτε ἔκ του
iib) οὔτε ἄλλο ἐξ ἐκείνης γενήσεται,
iii) εἴπερ ἐξ ἀρχῆς δεῖ τὰ πάντα γίγνεσθαι (245d4–6).

The consequent (iia & b) is to be denied, thus establishing the negation of the antecedent (i) by *modus tollens* (given that iii is axiomatically true). In

---

[23] Hackforth (2) 66n1.

[24] The expression is, I suggest, to be contrasted to the description, that occurs three times in *Laws* X, of the soul as τὴν αὐτὴν αὑτὴν δυναμένην κινεῖν (894d3–4; cf. b10, 896a1). The argument for the existence of soul requires only that soul is able to move itself; the argument for the soul's immortality requires that soul indeed does move itself.

[25] The ἀλλά (245c8), though likely having an explicative force, also seems to contrast ἑαυτό (c8) with ἄλλοις (c9).

[26] The identity of B and the ἀρχὴ κινήσεως is never proven in the Platonic corpus. Even in the *Laws* X the identity is assumed rather than demonstrated. In the *Phaedrus* the emphatic οὕτω δή (245d6–7) does not suggest that the identity is proven here. Rather the phrase ranges over the whole μέν . . . δέ contrast (d6–8). The sentence (d6–8) means: since it has been shown that the ἀρχὴ κινήσεως is neither generated nor perishable, and since B is an ἀρχὴ κινήσεως, then B, too, is ungenerated and imperishable.

[27] For the correct reading and interpretation of the sentence εἰ γὰρ . . . ἀρχὴ γίγνοιτο (d2–3), see Reed (2) 6. A consequence of this sentence is that "Plato assumes that ἀρχή cannot come to be from ἀρχή."

[28] The ensuing sentence ἐπειδὴ δὲ . . . ἀνάγκη εἶναι (d3–4) is troublesome because it seems to claim that the argument that B is imperishable will be a consequence of the argument that B is ungenerated, and yet the later argument (d4–6) is not cast in such a way that it employs the earlier one. I think though that d3–4 may only be claiming that the same axioms used to prove the ungeneratedness of B are also the axioms used to prove the imperishability of B.

this way, the desired conclusion is reached that the ἀρχὴ κινήσεως is imperishable. The denial of iia, entails the consequence that an ἀρχὴ κινήσεως (αὐτή, d5) comes to be out of something (ἔκ του, d5). This is surprising, since this very claim was seemingly denied at d2–3, where ἔκ του ἀρχὴ γίγνοιτο is denied. The ἔκ του of d2, however, seems to range only over things that are other than the ἀρχή at d5, namely, over α) material objects whose motions are described at c6 and β) other ἀρχαί.[29] The denial of iia does not seem to rule out that a principle could be said to cause its own motion. The reason for the denial of iia, therefore, I suggest, is the claim that B does not lose contact with itself but causes its own motion.[30] Here lies the significance of the claim that B is immortal insofar as it οὐκ ἀπολεῖπον ἑαυτό (c8).

To deny iib, it is not sufficient to make the identity between B and an ἀρχὴ κινήσεως or to assert that all motion comes to be from an ἀρχή, since there might not be any ἀρχὴ κινήσεως or any κίνησις. The lines d8–e2 entertain and dismiss these alternatives as impossible and so I suggest are an integral part of the argument for immortality, as giving the reason for the denial of iib. That the lines occur after the formal conclusion of the argument (d7–8) does not gainsay them this role and does not relegate them to the status of having been added merely for good measure to an argument that is complete without them, as most critics take them to be. In these lines (d8–e2), we are asked (implicitly) to entertain again, as at d4–5, the absence of an ἀρχὴ κινήσεως,[31] such that all the universe and all becoming would, by falling together, cease to move. However, if out of this calm arose spontaneously, without psychic causes, an acosmic flux of phenomena, the supposedly *per impossibile* consequence that becoming would never again have that from which it would become and move (e1–2) is completely robbed of any force, and so the argument that an ἀρχὴ κινήσεως is imperishable because only it can cause and account for the motions of other things (ἄλλο, d5) collapses. Our point II), then, succeeds in the case of the *Phaedrus*. The presence of non-psychically caused motions would destroy the argument for the immortality of soul.

---

[29] See n27 above.

[30] This is not to deny that B is ungenerated, for it neither implies that B comes into existence *simpliciter* nor that it comes to be what it is from some other earlier condition.

[31] Συμπεσοῦσαν and αὖθις (e1) make it clear that only the condition of the ἀρχή as having perished and not as not yet having been generated is under consideration in the hypothesis. This suggests that the logical role of e1–2 is not to add force to the conclusion οὔτε γίγνεσθαι δυνατόν (d8) as its position might suggest, but that it has some other role.

BODILY MOTION AND THE ARGUMENT FOR THE EXISTENCE OF SOUL IN THE *LAWS* X

I will now argue that in the *Laws* X, I) the motions covered by the ἀρχὴ κινήσεως doctrine would have to include the motions which characterize the acosmic flux of phenomena in the *Timaeus* and II) the presence of non-psychically caused motions would destroy the argument under consideration.

As in the *Phaedrus*, the motions which characterize the bodily as such in the *Laws* X are the same as those which we have seen characterize the pre-cosmic chaotic accompanying causes of the *Timaeus*: ἡ δι' ἕτερον ἐν ἄλλῳ γιγνομένη κίνησις [894e4–5, 895b6–7] . . . σώματος οὖσα ὄντως ἀψύχου μεταβολή (896b4–8). Moreover, the components of the pre-cosmic flux of the *Timaeus*, namely, γένεσις and φθορά[32] are said to fall under the scope of the ἀρχὴ κινήσεως doctrine in the *Laws* (894a2, 7; cf. 891e5). So point I) seems to succeed also in the case of the *Laws*.

Further, the existence of a pre-cosmos of chaotic motions would destroy the argument for the existence of soul. This argument has two parts (894e4–895a3 and 895a5–b3), which closely parallel each other and each of which independently establishes the existence of that which moves itself. The second is an elaboration of the first.

The first argument is a *reductio*. We are asked to assume that the only motions which exist are those which are moved by others and which in turn move still other motions. The absurdity follows that if everything requires something else to move it, there would never be a first motion and so (implicitly) there would be no motion at all, which is contrary to fact. Suppose, however, that there always had been chaotic physical motions as in the pre-cosmos of the *Timaeus*. This would impose two equally disastrous alternatives upon the argument. It would mean either that in the sequence of causes of motions characterized by "A" reaching into the past, there never was a first term to the sequence[33] or that material objects which have motions of type A move, nevertheless, spontaneously such that the extension of motions of type A and type B are the same. Either alternative does away with the need to establish a motion of type B over and above motions of type A.

---

[32] Τὸ γιγόμενον καὶ ἀπολλύμενον, *Timaeus* 28a3 = τὸ γιγνόμενον, 27d6, 50c7–d1 = (simply) the γένεσις (52d3) which is said to exist even before the universe came into existence (52d4) and which, as a result, is equivalent to the pre-cosmic flux of phenomena.

[33] It is the rejection of the argument's assumption that every series must have a first term that leads the followers of Hume to reject the validity of this form of the cosmological argument Cf. Meldrum 70n38.

The first part of the argument (894e4–895a3) attempted to show the absurdity of assuming the existence of motions only of type A. The second part (895a5–b3) asks us to assume that not even motions of type A exist, so that no motions of any kind exist. It asks us to assume that all is at rest. It is then assumed that some motion must arise; the only question is what kind of motion would arise first. A Newtonian would, of course, wish to deny that if everything were at rest, any motion would ever arise. Plato, however, wishes to claim that since there is *ex hypothesi* no motion or shifting, the first motion to arise must necessarily be a self-motion (i.e., motion of type B). However, the very hypothesis is ruled out of court if one presupposes a pre-cosmic chaos which has always existed. Moreover, if chaos has always existed exactly because matter moves spontaneously, then even if the motion of matter was momentarily arrested, it would commence again on its own once the arresting impediment was removed. So again the presence of a pre-cosmic non-psychically caused flux would destroy the argument and so point II) succeeds in the case of the Laws.

Vlastos, however, would wish to deny both I) and II) of the *Laws*. Against I), he claims that "at face value, this [*Laws* 896b] asserts that the soul is itself the cause of the instability of becoming; that apart from soul reality would be untroubled by transience. But this is un-Platonic."[34] Such an appeal to what is or is not Platonic is odd in Vlastos' case, since he finds Plato neither coherent[35] nor even consistent on such major issues as whether or not that which moves itself is ungenerated.[36] But in any case, Vlastos goes on to conclude that "the one thing he [Plato] cannot mean in the *Laws* is that soul is the source of Heraclitean flux."[37] The only reason Vlastos cites for this, however, is that in the *Phaedo* (79e) it is claimed that soul is in every way more like being than becoming. However, the only commitment made in this passage of the *Phaedo* is that soul, like the Ideas, is non-material, and this property for Plato hardly rules out soul as a cause of becoming.

Against II), Vlastos claims not only that a pre-cosmic non-psychically caused flux would not destroy the argument of *Laws* X, but is actually presupposed by it: "Γένεσις must be presupposed. It must be 'there', before soul can supervene to 'rule' it. But if it is 'there', it must involve mo-

---

34 Vlastos (2) 398. Vlastos claims that the assertion of the universal standstill (*Laws* 895a–b) is enemy territory and that the supposition of a standstill in the *Phaedrus* is the apodosis of a *per impossibile* hypothesis. Vlastos (2) 398n1. However, *Phaedrus* 245d4–6 is decisive against Vlastos' position, for it clearly states that if there is no ἀρχὴ κινήσεως, there is no motion of any kind.

35 Vlastos (2) 399.

36 Vlastos (2) 397n3, cf. (3) 414n1. For another objection to this stance of Vlastos, see Clegg 59.

37 Vlastos (2) 398.

tions of some sort."[38] However, it is not a Platonic principle that the concept 'ruled' (read 'ordered', for strictly speaking only souls can be ruled) must entail that that which is ruled or ordered is in motion.[39] One can, of course, order things that are at rest.[40] It happens that in the *Timaeus* and *Statesman* myth that which is ruled is in motion, but it is so independently of the fact that it is eventually ordered or ruled.

I take I) and II), then, as established for both the *Phaedrus* and the *Laws*.

I now turn to Easterling's argumentation for the limitation of the ἀρχὴ κινήσεως doctrine to the created world. I treat his as a special case not because his mode of argumentation is basically different than the critics discussed so far,[41] but because he goes on at some length to try to show that this view coheres with a fair number of texts. Easterling's prime text is *Laws* 896e8–897b4, which he claims at least hints at the limitation of the ἀρχὴ κινήσεως doctrine to the created world and at the possibility of non-psychically caused pre-cosmic motions within the *Laws* itself.[42] Easterling is quite right to point out that the key word παραλαμβάνουσαι (897a5) means "to take over" something which is already in existence.[43]

---

[38] Vlastos (2) 398.

[39] See further n46 below.

[40] Indeed some of the things that lie ready to hand for the soul's ordering or ruling are explicitly stated to be at rest (893c1); the ensuing sentence (893c1–2), especially when compared to *Timaeus* 52a3, c6–d1, makes it clear that the objects at rest here are phenomenal objects and are not (as they are sometimes assumed to be) Platonic Ideas.

[41] Easterling's argument ultimately, like Vlastos', Hackforth's, and Robinson's, rests on the argument from silence. Easterling 38. Further, like Robinson, Easterling emphasizes the differences between the cosmos and the pre-cosmos: "The created universe is an essentially different place [than the pre-cosmos]." Easterling 37. However, this 'essential difference' turns out to be on Easterling's account only that soul causes motion in the created world but not in the pre-cosmos. Easterling 37–38. So this essential difference turns out, in Easterling's case, simply to be a restatement of what he is trying to establish, and adds no support to the argument from silence.

[42] Easterling 32–33.

[43] Easterling 31. I do not, however, think the term here is being used in the sense of "to take to oneself as a partner, auxiliary or ally" (see *LSJ*, s.v. παραλαμβάνω, sense II, 1), since it is not that which is taken over, but that condition into which that which is taken over is brought that is treated instrumentally: ἄγουσι πάντα εἰς αὔξησιν καὶ . . . καὶ πᾶσιν οἷς ψυχὴ χρωμένη (897a5–6, b1).

Further, therefore, though Easterling is right to point out that δευτερουργοί (a4) "does not mean secondary in the sense of 'derivative' or 'induced', i.e., 'derived from something primary' "but means "'producing an effect that is secondary'," he is wrong to interpret 'producing secondary effects' as being equivalent to 'subordinate or subservient to something primary'. Easterling 31. It is true that

Easterling, however, goes beyond the text when he takes this passage as denying that the motions already in existence (when taken over by soul) are not also caused by soul.[44] For the contrast between the πρωτουργοί and the δευτερουργοί motions reintroduces the distinction between that which, by moving itself, can initiate motions in others (B) (894e8) and that which cannot move itself and which can only transmit motion (A) (894e4–5). And B is unequivocally said to cause A (892a, 894e–895b, 896d, 966d–e, *Phaedrus* 245c) as even Easterling admits, but seems to forget immediately upon stating.[45] But even more decisive against Easterling's interpretation of 896e–897b is that his interpretation of this passage is exactly the position of the materialist against whom Plato is arguing at 889a. It is Plato's *opponents* that argue that art (τέχνη) takes over (λαμβάνουσαν) becoming (γένεσιν = πάντα τὰ πράγματα γιγνόμενα καὶ γενόμενα καὶ γενησόμενα, 888e4–5) from nature (παρὰ φύσεως), and itself molds and shapes all things (889a5–7). It is exactly against this view that Plato advances the ἀρχὴ κινήσεως doctrine. This view, then, cannot possibly be what is expressed as doctrine at 896e–897b. Easterling in a desperate attempt to reconcile the *Timaeus* and *Laws* X reads the pre-cosmos of the *Timaeus* into the *Laws*, but only at the expense of making the *Laws* contradict itself.[46]

Easterling has equal trouble reading his version of the ἀρχὴ κινήσεως doctrine into the *Timaeus*. He wishes to advance an interpretation which will be compatible with either a literal or a non-literal reading of the creation story, but has difficulty both ways. If, as he does, Easterling accepts

---

psychic agents *can use* the δευτερουργοὶ κινήσεις *as* subservient instruments exactly because the δευτερουργοὶ κινήσεις do affect each other, but it is not part of their nature *per se* to be subservient.

44 Easterling 31.

45 Easterling 26.

46 This argument against Easterling, based on the obvious sense of *Laws* 889a, holds equally well against Vlastos' view (see above) that γένεσις must be presupposed as existing in *Laws* X "before soul can supervene to 'rule' it." Vlastos (2) 398. Clegg is correct to draw attention to the fact that *Laws* 889a contradicts the repeated claims in the *Timaeus* which at face value assert that the Demiurge takes over becoming (as given prior in nature) and molds and shapes it. Clegg 54. Clegg, however, simply begs the question of the unity of Plato's thought on psycho-physical matters by going on to claim that therefore such statements in the *Timaeus* cannot be taken at face value. The materialists which Plato attacks at *Timaeus* 46c–e need not be co-extensive with the materialists attacked at *Laws* 889a. Indeed the nature of the attack in each case is different, since the claim that *not all* motions are physical motions (*Timaeus* 46c–e) is different than the claim that *all* motions have a non-physical origin (*Laws* 889a *et passim*). There is no reason, therefore, why the materialists of *Laws* 889a could not include the author of the *Timaeus*.

Cherniss' interpretation of 58a–c, which entails demiurgic activity of the World-Soul,[47] he is pretty well committed to a non-literal reading of the creation story, since 58a–c is explicitly meant to explain the maintenance of the flux of the phenomena, given the conditions of the pre-cosmos described at 52d–53a (see 58a1–4).[48] Easterling cannot claim that 52d–53a describes a pre-cosmic state and yet that 58a–c describes a cosmic state; both passages must describe the same state whichever it is. But if Easterling opts for a non-literal reading of the creation myth, he is hard put to say what non-mythical factor in the created world the pre-cosmos is to represent, since he wishes to claim that all motions in the created world (which on the non-literal reading of the myth is the only world there is) are caused by soul and yet he asserts "the independent status of the pre-cosmic motions" of the *Timaeus*.[49]

The best Easterling can do in this strait is to more or less deny that the pre-cosmos represents actual motions. He does this in two ways. First, he seems to claim the description of chaotic motions is merely a symbol of disorder and is not meant to be taken literally as describing motions: "Just as reason and order can only be manifested in purposive orderly movement, so by contrast the opposite qualities are most *effectively displayed* in random, disorderly movement."[50] That Plato in the *Timaeus* meant the descriptions of the phenomena in flux to be read literally is evident from his contrasting this condition of the phenomena with the condition of the phenomena as images, which are disorderly strewn across the receptacle (49b–50b, on which see chapters 3 and 4 above). The one condition, therefore, cannot be taken as a symbol of the other. That Plato believed that the flux of phenomena was meant to be taken literally should also be clear from *Philebus* 59a–b. Further though, it is unusual for an author to try to 'save' a symbol's details which are irrelevant, even contradictory, to that which the symbol symbolizes: if the description of the pre-cosmic flux is simply an effective way of displaying disorder and is not meant literally to be a description of a kind of motion, it is more than strange that Plato should install in the *Timaeus* the elaborate mechanism of the primary corpuscles as an attempt to save the motions of the pre-cosmos (*Timaeus* 56a–b with 58a–c; note διασῳζομένη τὴν ἀεὶ κίνησιν, 58c3).

Second, Easterling, in trying to read the motions of the pre-cosmos as representing a static condition, puts a great deal of weight on the description of the pre-cosmic receptacle as being filled with δυνάμεις (52e2). Easterling takes δύναμις in a nearly Aristotelian sense to indicate

---

47 Easterling 37 with n28. Cherniss (1) 444–50, especially 448–50 with notes.

48 Cf. Tarán n131.

49 Easterling 26.

50 Easterling 37, my emphasis.

"a capacity that naturally leads to a κίνησις . . . only when actualized by an external agent capable of initiating motion (i.e., a ψυχή)."[51] Unfortunately for Easterling's position, the two other occurrences of δύναμις (33a, 56c) which he cites as support for his position actually torpedo it, since in both cases the δυνάμεις clearly include *actual* motions (n.b. προσπίπτοντα λύει, 33a4–5; τὰς κινήσεις καὶ τὰς ἄλλας δυνάμεις, 56c4).

Easterling has equal difficulty trying to read his version of the ἀρχὴ κινήσεως doctrine into the *Timaeus*, when the *Timaeus* myth is read literally. First, Easterling must claim that the δυνάμεις, though requiring actualizing agents to become κινήσεις in the created world, do not need such agency in the pre-cosmos in order to become κινήσεις.[52] Unfortunately for this view, δύναμις is being used, as Easterling even seems to admit, univocally in describing both the cosmos (33a, 56c) and the pre-cosmos (52e). In addition, for Easterling to admit actual pre-cosmic motions and yet to claim that all motions in the created world are initiated by psychic agents, he must assert that pre-cosmic motions are brought to a halt and are not carried over as constituents of the created world. Vlastos, Hackforth and Robinson are also, it would seem, committed to this claim, but only Easterling is explicit about it: "Once ψυχή is created that state of affairs [the pre-cosmic flux] comes to an end."[53] However, as I have suggested, the recurrence of acosmic motions attacking and ultimately overwhelming the World-Soul of the *Statesman* myth (a text to which Easterling makes no reference) and the parallel attack of the infant soul by the flux of phenomena in the *Timaeus* (43a–b, especially a5–b2), both events in the created world (i.e., the world in which there are souls), strongly suggests that the pre-cosmic flux continues to be a dynamic factor within the ordered universe, as do also the references throughout the corpus to the phenomenal flux (e.g., *Philebus* 59a–b, *Cratylus* 439d) which do not distinguish between cosmic and acosmic periods.

Easterling, like Vlastos, Hackforth, and Robinson, is not successful in interpreting either the *Timaeus* or the *Laws* and *Phaedrus* as supporting, or even as being consistent with, the view that the scope of the ἀρχὴ κινήσεως doctrine is meant to be limited to the ordered universe and is not meant to cover the motions of the pre-cosmos. It does not seem possible then to square the *Timaeus* and *Statesman* with the *Phaedrus* and *Laws* X insofar as the former texts view the corporeal as moving spontaneously and as constituting on its own a positive source of evil and the latter texts view all motions and all positive evils as having psychic origins.

---

[51] Easterling 35.
[52] Easterling 37–38.
[53] Easterling 38.

NINE

The World-Soul in the Platonic Cosmology

(*Statesman, Philebus, Timaeus*)*

In each of his major cosmological works, the *Timaeus*, the *Statesman* myth, and the *Philebus* (26e–30e), Plato asserts that the body of the whole universe is alive and possesses a single World-Soul which extends throughout it. I wish to offer a new interpretation of the role of the World-Soul, an interpretation which gives the World-Soul a special function in the economy of the Platonic cosmology and which explains why Plato would place such repeated emphasis on the existence of such an odd-sounding creature. I suggest that Plato is not viewing the World-Soul on the model of the *Phaedrus* and *Laws* X, which view soul as self-and-other moving motion. Nor, I suggest, is Plato viewing the World-Soul on the model of soul taken as a crafting agent that initiates order. Rather I suggest that Plato views the World-Soul merely as a *maintainer* of order against a natural tendency of the corporeal to be chaotic.[1]

It is important to notice that in each of these three cosmological dialogues Plato claims that the ordered World-Soul and the *order* of the World-Body are severally and in their synchronizations the products of the workings of a single, eternal, divine, rational Demiurge, which resides outside the universe.[2] Further, in all three of these dialogues the phenomena are viewed as necessarily in flux. The erratic flux of the phenomena wholly characterizes the pre-cosmic and acosmic periods of the *Timaeus* and *Statesman* myth, but in addition it remains a potent and considerable factor even within the ordered and ensouled cosmos (*Timaeus* 43a–b, *Statesman* 273c–d and add *Philebus* 43a, 59a–b and *Cratylus* 439d, which do not distinguish between cosmic and acosmic periods).

---

* Reprinted from *Illinois Classical Studies* 7 (1982) by permission of the editor and Scholars Press.

[1] In this chapter I do not discuss the role of the World-Soul in the *Timaeus* as an important object of human cognition. For that epistemological function of the World-Soul, see chapter 1, sections III and VIII.

[2] *Statesman*, 269c–270a, 272e–273e; *Philebus* 28d, 30a–d; *Timaeus* 31b-36e. I take these texts to be doctrinally consistent with each other.

With these observations in tow, I suggest that the World-Soul operates in the Platonic cosmology rather like a governor on a steam engine: the governor regulates the motions of the machine in such a way that the machine's self-sustained and independently originated motions, which owing to unpredictable conditions of combustion tend to run off to excess, are nonetheless uniformly maintained and do not destroy the machine itself. However, the governor neither initiates the motions it regulates nor is it itself the cause of its being synchronized with the machine. This synchronization, which enables the governor to govern, is derived from some external source. And like a machine-governor, the World-Soul is capable of maintaining order only within a certain range of natural disruptions (*Statesman* 273d–e).

If, as is the case, Plato believes material objects necessarily tend toward chaotic flux even in the formed and ensouled cosmos (*Timaeus* 43a–b, *Statesman* 273c–d), then it is natural that he should view one of the major functions of soul to be the maintenance of order against the natural tendency of the corporeal to be chaotic, thus saving the appearance of the continuous order which we indeed do observe in the phenomenal realm. For Plato the homeostatic conditions of the observed world cannot be explained *by* physical theories; rather, they have to be explained *in spite of* physical theories (of the sort articulated at *Timaeus* 58a–c).

World-Soul in the Statesman. In the *Statesman* myth the Demiurge is said to make the World-Soul, to make it rational (269d1), and to form the ordered World-Body (269d7–9, 273b6–7, 273e3). In the structure of the dialogue as a whole, the Demiurge is functionally contrasted with the World-Soul as a shepherd is contrasted with a human statesman (275a–276b). The Demiurge is like a shepherd in that he constructs or prepares on his own every material component of the objects of his craft, whereas the World-Soul is like a human statesman just to the extent that all the necessary material preparations and organizations of the object of its activities are handed over to it from sources other than itself. Since this distinction is asserted at the level of discourse of the divisions that make up the bulk of the dialogue, it is difficult to claim that the division of the Demiurge from the World-Soul is merely a literary exigency of the myth (as some critics claim is the case in the *Timaeus*).[3] In the case of the World-Soul, it is the whole order of the world that is handed over to it (273a7–b2). This is dramatically represented by the withdrawal of the Demiurge from the world, which he leaves in the care and control of the World-Soul. The World-Soul, though, unlike the human statesman, does not further organize the organization handed over to it (305e ff.). Rather the World-Soul simply tries to maintain the orderly homeostatic condi-

---

[3] The overwhelming tendency in Platonic scholarship has been to read Plato's claims about the Demiurge non-literally. See chapter 1, n33.

tions of the World-Body (as they are inherited from the Demiurge) against the necessary, erratic, even explosive (cf. ἐξανθεῖ, 273d1) incursions of the bodily, which tend to throw the organization of the World-Body *and* World-Soul out of kilter (273b, d). The World-Soul performs this task not by initiating order, but merely by trying to remember and preserve the orderings given from the Demiurge (273b1–2, c6). Eventually though, the bodily incursions succeed in disrupting the World-Soul's memory and the World-Soul thereupon loses its ability to maintain order. This decay necessitates the reappearance of the Demiurge to restore order to both the World-Soul and World-Body (273d–e). It seems then that the World-Soul is not being viewed as an initiator of orderly motion either in itself or in the World-Body. For this role is reserved for the Demiurge.[4]

If, in addition to regulating motions, the World-Soul were able to initiate new motion, it is not clear why the World-Soul must succumb to the disruptions of the corporeal. If it were able to induce new motion, it would be able not merely to keep a lid on disruptive forces but also to counteract and diffuse the cause of disruption. Further, if it had self-initiated thought and reason, items in the catalogue of self-motions in *Laws* X (897a–b), and did not have its rationality derived entirely from an external source, then it is not clear why, on its own, its failures of memory are irreparable and irreversible, such that it is necessary for the Demiurge to reappear and initiate new order.

It is, then, I suggest, to the homeostatic conditions of living creatures, rather than to their ability to self-initiate locomotion and to move other objects, to which Plato in the *Statesman* myth is primarily appealing, when he posits the world as a living creature (ζῷον, 269d1). This sort of appeal should be contrasted with the doctrine of the autokinetic soul in the *Phaedrus* and *Laws* X. In each of these texts it is not to homeostatic conditions of living organisms, but rather to the motor powers of living bodies to which Plato appeals in order to identify that which is autokinetic with soul. It is because living bodies move themselves and other things that we know that that which is autokinetic is soul (*Phaedrus* 245e, *Laws* 895c).[5]

World-Soul in the Philebus. The function of the World-Soul as a maintainer of homeostatic conditions, a function which results from its vivifying effects on bodies, is also evident in the *Philebus* (30a-c), a dialogue in which, as in the *Statesman* and *Timaeus*, flux characterizes the phenomena

---

[4] Nor is the World-Soul the source (either direct or indirect) of the disorderly motions which erupt into it. See chapter 7.

[5] Robinson (mistakenly, I think) takes several reflexive phrases in the myth as referring to autokinetic soul. Robinson (1) 134, 135, 139. For discussion, see chapter 7, n3.

(43a, 59a–b) and in which there is not the slightest trace of the autokinetic doctrine. As in the *Statesman* and *Timaeus,* the order of the world's body is derived from the transcendent rational Demiurge (28d). In addition, as also in the other two dialogues, the Demiurge is the cause of the presence of the World-Soul and its rationality in the ordered World-Body (30c–d). And so again there is no suggestion that the World-Soul is the efficient cause of the motion or the order of the World-Body.

Rather, the World-Soul here is viewed as standing to the order of the universe (as represented in the orderly years, seasons and months) as our souls stand to our bodily order (as represented in health) (30b–c). The *only* actions of our soul-body complex here mentioned as being relevantly paralleled in the World-Soul/World-Body complex are physical exercise and (self-)doctoring, both of which maintain or restore from deviation the homeostatic condition of the body. Significantly, nutritional and sheltering arts are not on the list of parallel practices, since, we may assume, they both involve manipulations of the external world, while the actions relevant to the World-Soul are internally directed. (For exercise and proper doctoring dealing only with the relation of the body with itself and not with the external world, see *Timaeus* 89a–b.) Our souls are the cause of the maintenance of health or proper orderings of our bodies against a natural propensity towards disease, which is viewed as a sort of internal corrosion (cf. *Timaeus* 82a–83a). Analogously, the only actions entertained as being performed by the World-Soul are those regulations of the World-Body which maintain its order against disruption natural to it. Indeed it is to save the appearance of rational order that it is claimed that there must be a World-Soul (30c–d). Plato does call the World-Soul a cause (30a10), but Plato is only committed to viewing it as a cause in the sense of an agency of sameness, of maintenance.

World-Soul in the Timaeus. In the *Timaeus* we are told little of the nature of the functional relations between the World-Body and the World-Soul. All that we are told is that the World-Soul is the mistress and governor (δεσπότιν καὶ ἄρξουσαν, 34c5) of the World-Body. What form this governance is to take, we are not told. I suggest, though, that it entails no more than the sort of governance I have already mentioned, namely, the maintenance of order. There is no suggestion in the *Timaeus* that the World-Soul is either autokinetic[6] or is the efficient cause of either the order or the motion of the World-Body. The form and orderly motion of both the World-Soul and World-Body are derived from the Demiurge (31b–36e). In commenting on 36d–e, Cornford writes: "The above sentences reiterate the emphasis already laid at 34b on the fact that the soul extends throughout the body of the world from center to circumference, and

[6] On *Timaeus* 37b5, which is sometimes seen as such a suggestion, see Cornford's note *ad loc.,* which has not been superseded by later discussions.

communicates its motion to the whole."[7] Now it is true that the World-Soul is so extended, but there is not a word in the text about the World-Soul communicating its motion to the World-Body. Rather we have in the text a highly detailed account of parallel structures and synchronized motions (as represented in celestial dynamics) between the World-Soul and World-Body. This synchronization is derived from the Demiurge and is not of the World-Soul's making. Plato is free to have said otherwise. For, when at 34c Plato admits that his narrative order was mistaken and misleading in having spoken of the World-Body being composed prior to the World-Soul, he could have taken the opportunity to claim that it was merely an exigency of his narrative order that forced him to claim that the Demiurge rather than the World-Soul composed the order of the World-Body and initiated its orderly motions, since in the mistaken narrative order the World-Soul did not even exist when the World-Body was established. But later, the mistake in narrative order having been pointed out, the cosmological claims of 34b are allowed to stand and are reiterated: the structure and motion of the World-Soul and World-Body severally and the synchronizations between them are all workings of the Demiurge (36d–e). Taken at face value, the *Timaeus* strongly suggests that when Plato claims governance on behalf of the World-Soul, he does not mean that the World-Soul acts as a crafting agent or as an efficient cause of motion.

At first inspection, the World-Soul strikes us as perhaps the oddest of many odd components of Plato's cosmology in that it is highly counter-intuitive: the world just does not feel like an animal. Most of it is clearly inert. Further, the World-Soul appears to be redundant or useless ontological baggage on most interpretations, which assimilate it either to the autokinetic doctrine or to the view of soul as a crafting agent. For if the World-Soul is merely one more autokinetic soul, it has no special function in the economy of Plato's cosmology. And similarly, if the World-Soul is viewed (incorrectly in my opinion) as mainly an agent that crafts external objects, then it becomes indistinguishable in function from the Demiurge. If it is understood, though, that Plato views order among the phenomena as the thinnest of veneers, made out of and extending over that which is inherently rotting, we then see that it is reasonable for Plato: 1) to assume the existence of a regulating agency which, on the one hand, is necessarily non-material, but on the other hand, is immanent in the corporeal world, thus explaining the persistence of what sensible order there is in the world, and 2) to leave the original source of the order of both the World-Body and World-Soul outside the soul-body complex, unaffected by the natural corrupting influence of the corporeal.

---

[7] Cornford (2) 93.

If my interpretation of the World-Soul is correct, two additional oddities of its characterization are explained. First, one typical function of an ensouled rational creature is deliberation and practical reason. Second, typically for Plato souls are viewed as capable of discarnate existence. Yet neither of these characterizations holds of the World-Soul in the *Timaeus, Statesman,* and *Philebus.*

First, though in the *Timaeus* the World-Soul has true opinion and contemplative reason (37b–c, especially c1, 2), we never hear here or in the *Philebus* or *Statesman* of the World-Soul deliberating or making decisions, as do the Demiurge and statesman (*Statesman* 305e ff.; *Republic* 484c–d, 500e). The World-Soul's rationality is not that of planning or producing with the aid of paradigms, as is in large part the rationality of the Demiurge and statesman. But, if as I have suggested the World-Soul's function is that of maintaining order rather than initiating order, this is to be expected.

Second, unlike personal souls, the World-Soul is never viewed as existing in a discarnate condition. If its function is the maintenance of homeostatic conditions of material objects, it can only do this by being present in them. Insofar as the ordered world is to exist sempiternally (*Timaeus* 38b–c), so too must the World-Soul abide in it.

Aristotle thought that through the whole of the natural world the motions of bodies on their own were constant, uniform, and orderly enough that it made sense to describe both animate and inanimate objects as moving homeostatically, as though the whole of nature were like someone who heals himself (*Physics* II, 8, 199b30–32). Plato felt that the corporeal just by itself was so chaotic that *at best* ensouled objects could maintain orderly homeostatic conditions, and even then with only limited success.

Further, though the World-Soul for Plato, *in order to* have its special function as a maintainer of order, is necessarily immanent in the corporeal, it is not immanent in the corporeal as the result of its ontological status or make up, which is the same as that of human souls, which *are* capable of discarnate existence (*Timaeus* 41d–e). *In principle,* then, the World-Soul should be capable of discarnate existence. So that though *in fact* the World-Soul is immanent in the material world, it is not to be thought of as merely the functioning or actualization of a body of a certain type, as is the soul for Aristotle, which as such is not capable, even in principle, of discarnate existence (*De Anima* II, 1, especially 413a4). There are additional reasons to suppose the World-Soul is not the actualization of a body of a certain type. One, the World-Soul is a precondition for any matter even being the sort of thing which might be ordered enough to be considered an organ with a function. Two, the World-Soul is not the functioning of a body, but is that which makes it possible that the functions of various bodily parts are sustained. And three, unlike Aristotelian souls, the World-Soul has no limit on what sorts of bodies it may vivify. There is no proper matter for the World-Soul: it extends through all of the primary bodies (earth, air, fire, water). The World-Soul, unlike Aristotelian soul, is self-substantial independently of its material inherence. The immanent World-Soul is not, then, a step in the direction of either *Laws* X or *De Anima* II.

TEN

The Relation of Reason to Soul in the Platonic Cosmology: *Sophist* 248e–249d*

Since Cherniss' *Aristotle's Criticism of Plato and the Academy* I, there has been nearly universal agreement among critics that Plato's God or divine Demiurge is a soul.[1] Yet the *prima facie* evidence is that the Demiurge is not. In all three of Plato's major cosmological works the *Timaeus,* the *Statesman* myth, and the *Philebus* (26e–30e), the Demiurge is fairly extensively described and yet not once is he described as a soul. Rather souls, and especially the World-Soul and what rationality souls have are viewed as *products* of the Demiurge (*Timaeus* 35a, 36d–e, *Philebus* 30c–d, *Statesman* 269c–d).

Nonetheless, the overwhelming critical opinion is that since the demiurgic God of these works is described as rational, *this* entails that God is a soul. Three texts are adduced to prove this: *Timaeus* 30b3, *Philebus* 30c9–10, and *Sophist* 249a. These texts are taken as claiming A) that if a thing is rational, then it is a soul. Proclus saw that at least the *Timaeus* passage can mean only B) that when reason is *in something,* what it is in must be an ensouled thing. The rhetoric of the *Timaeus* sentence strongly suggests that reading B is correct[2] and the argumentative context of the

* Reprinted from *Apeiron* 16 (1982) by permission of the editors.

[1] Cherniss (1) appendix XI, which is in part an attack on Hackforth (1). Cherniss' view on this point has been accepted by a wide range of critics: Tarán n34; Lee (4) 89; Robinson (1) 68, 102, 114, 142; Vlastos (3) 407. Guthrie, though, maintains Hackforth's position that the Demiurge is not a soul, 215, 275n1, as does Menn 10–13, 19–24.

[2] The argument at *Timaeus* 30a–b would go through if Plato simply said "all reason is soul or is ensouled." There is nothing here or elsewhere in the corpus to keep him from saying this, if this is what he meant. And this may have been what he meant here if he had said only νοῦν χωρὶς ψυχῆς ἀδύνατον [εἶναι], but in fact what he says is ἀδύνατον παραγενέσθαι τῳ. The emphatic position of τῳ, the use of γίγνομαι rather than εἰμί (in a passage which is critically concerned with distinguishing the terms, cf. especially 27d5–28c3 and n.b. γενέσθαι, 30c1) and the use of the prefix παρα-, all look like conspicuous literary consumption in an otherwise highly economical passage, *if* they do not affect the sense of the sentence. The sentence, then, seems to be the articulation of a general metaphysical principle that when reason comes to be in anything whatsoever, that in which it

Philebus sentence (properly understood) requires sense B.[3] This leaves, as Cherniss is willing to admit,[4] the *Sophist* passage alone as bearing the whole weight of Plato's alleged commitment to the view A) that everything that is rational is a soul. I wish to give a new, tentative interpretation to this passage which shows that it is, like the *Timaeus* and *Philebus,* committed only to the weaker claim B), that when reason is *in something,* it is so along with soul. This leaves the Demiurge who is not in anything free to be rational without being a soul and to serve rather as a maker of souls.

The *Sophist* passage runs:

> Shall we let ourselves easily be persuaded that motion and life and soul and mind are really not present in complete being, that it neither lives nor thinks, but awful and holy, devoid of mind, is fixed and immovable?
> —That would be a shocking admission to make.
> But shall we say that it has mind, but not life?
> —How can we?
> But do we say that both of these exist in it, and yet go on to say that it does not possess them in a soul?
> —But how else can it possess them?
> Then shall we say that it has mind and life and soul, but, although endowed with soul, is absolutely immovable?
> —All those things seem to me absurd.
> And it must be conceded that motion and that which is moved exist.
> —Of course.

---

comes to be is an ensouled thing. The principle comes into play in the case at hand in the *Timaeus* because Plato is trying to establish "the nature of the visible" as "possessing reason" (30b1, 2). However, the principle is not said to hold exclusively of material objects, contra Hackforth. Hackforth (1) 445. *Timaeus* 46d5-6, which is sometimes viewed as committing Plato to the view that all reason is in soul, should be interpreted in light of 30b.

[3] As in the *Timaeus* passage, the use of γίγνεσθαι (*Philebus* 30c10) and ἐγγίγνεσθαι (d2) rather than εἶναι is probably not accidental. The argument of which the claim "reason cannot come to be apart from soul" (c9–10) in part runs as follows. Premise 1) there are in the World-Body traces of rationality represented by the orderly years, seasons, and months. This rationality in the *Timaeus* is called "the circuits of intelligence in the heavens" (47b7, cf. 34a). Reason here is a qualification of the bodily rather than a qualification of soul. Premise 2) just as our bodies naturally are given over to disorder (represented by disease) so the whole universe left on its own is in chaotic flux. Conclusion: therefore, the World-Body requires a soul whose function is to maintain the homeostatic conditions of the rational, orderly motions of the bodily against the natural propensity to disorder. Soul in this sense is necessary for reason in this sense. Therefore, the argument requires that 30c9–10 mean that when reason is in something, it is so in an ensouled thing and not that reason is necessarily a qualification of soul.

[4] Cherniss (1) 606.

> Then the result is that if there is no motion, there is no mind in anyone about anything anywhere.
> —Exactly.
> And on the other hand, if we admit that all things are in flux and motion, we shall remove mind itself from the number of existing things by this theory also.
> —How so?
> Do you think that sameness of quality or nature or relations could ever come into existence without the state of rest?
> —Not at all.
> What then? Without these can you see how mind could exist or come into existence anywhere?
> —By no means.
> And yet we certainly must contend by every argument against him who does away with knowledge or reason or mind and then makes any dogmatic assertion about anything.
> —Certainly.
> Then the philosopher, who pays the highest honor to these things, must necessarily, as it seems, because of them refuse to accept the theory of those who say the whole is at rest, whether as a unity or in many forms, and must also refuse utterly to listen to those who say that being is universal motion; he must quote the children's prayer, "all things immovable and in motion," and must say that being and the whole consist of both (248e6–249d4, Fowler adapted).

The context is important. Plato has just wrested from the materialists the concession that at least some small part of reality is non-material (τι καὶ σμικρόν, 247c9–d1). He then gives two arguments to show complementally that the friends of the Ideas are committed to the view that at least some small part of what really is in motion, that the sum total of what is cannot consist merely of the immutable objects of reason, on the one hand, and the phenomenal flux, on the other (248a).[5] The first argument is difficult and much debated (248a-e).[6] The second is the passage just quoted (248e–249d). For the purpose at hand, one small but crucial point is made in the course of the first argument: when reason is viewed as participation in what really is, the only reason here being discussed is *our* reason achieved by *our* souls through thought (ἡμᾶς κοινωνεῖν διὰ λογισμοῦ ψυχῇ πρὸς τὴν ὄντως οὐσίαν, 248a10–11).

The second argument (248e6 ff.) begins with a shift from talking of what really is to talking of what completely is (e7). Now apparently what

[5] When in contrasting the phenomenal world to the eternally immutable, Plato describes the phenomenal world as γένεσιν ἄλλοτε ἄλλως (248a12–13) or simply γένεσιν, he is asserting that the phenomenal world is constantly in flux, not merely that it is capable of change (cf. *Philebus* 59a7–b1).

[6] For widely varying interpretations of the first argument, see Cherniss (7) 311–12, Owen (2) 336–40, Keyt (2).

counts as part of what completely is will not be anything that is in flux; otherwise, of course, there would be no need to establish motion as part of what completely is—one could simply point to the phenomenal flux.[7] The argument to show that there is motion in what completely is has two prongs. First, it is claimed that our reason entails soul, which in turn entails motion, and so is different than what immutably is. Second, it is claimed that our rational soul cannot be in flux, and so also is necessarily distinct from the phenomena. Jointly the two prongs establish motion as part of what completely is. The steps of the argument run, I suggest, as follows:

1) when mind exists or comes to be in something anywhere at all (παρεῖναι, 249a1; νοῦν ἔχον, a2; ἐνόντ', a6; μηδενί, b6; ὁπουοῦν, c4), then that thing is ensouled.[8]
2) soul entails some sort of motion (249a10).
3) if, though, this motion is merely flux, we have not really shown that there is something other than the immutable and that which is in flux (249b8–10).
4) the motion of rational soul is uniform and not in flux (249b12–c1).
5) therefore, rational soul, if it exists, is different both from what immutably is and from that which is in flux.
6) but, reason *is* in us (those who deny this contradict themselves) (249c6–8 with 248a10–11).
7) therefore, something moving exists which is part of what complete is (249c10–d4).

We have in this short argument five indicators (step 1) that Plato is here talking not of mind *simpliciter*, but only of mind as it is possessed by something else or present in something else. And it is sufficient for the argument to succeed for Plato only to be claiming that if reason is present in something, soul too is present there, when he writes the crucial line "but, do we say that both of these [reason and life] exist in it [what completely is], and yet not go on to say that it does not possess them in a soul?" (249a6–7). The sentence need not be construed as claiming all reason dwells in soul. I think Cherniss is wrong to claim that "Plato could not have formulated [the argument at 249a] if he had believed that there is *any* real νοῦς which does not imply soul."[9] For, all that is re-

[7] Therefore τὸ παντελῶς ὄν cannot refer just to the phenomenal world as Hackforth would have it. Hackforth (1) 444–45. Nor does it refer exclusively to the Ideas as it does in the *Republic* and *Timaeus.*

[8] Μηδενί means "not in anyone" (so Fowler), as the objects of reason are covered by the phrase περὶ μηδενός (249b6). I take ὁπουοῦν as qualifying both ὄντα and γενόμενον at 249c3.

[9] Cherniss (1) 607.

quired for the argument to succeed is that a) there exists *some* soul and b) this soul is not chaotic. Both a) and b) are established by claiming that rationality exists *in us*.

Cherniss needs to construe the sentence the way he does, for he wants the second argument to end here (249a7) and wants the new contrasting point starting at 249b8 (καὶ μὴν . . . αὖ, step 3 on my construction) to constitute a needed rehabilitation of the (disastrous) first argument (248a–e), rather than to complete the second.[10] He construes this section (249b8 ff.) to mean that if the Ideas really move, as per the first argument, reason will be destroyed, for the objects of reason must remain immutable while only the soul moves.[11] Cherniss takes τὸ κατὰ ταὐτὰ καὶ ὡσαύτως καὶ περὶ τὸ αὐτό here (249b12) as referring to the Ideas, as a similar phrasing at 248a12 seems to. To so construe the argument, though, is to require that φερόμενα and κινούμενα (249b8) are thought to apply to the Ideas, that being known entails locomotion (φερόμενα), *as a result of* the first argument (248a–e). This is unlikely. Rather the phrase φερόμενα καὶ κινούμενα is a general description here for flux; it is equivalent to γένεσιν ἄλλοτε ἄλλως (248a12–13).

Flux here (249b8) is being entertained as a possible type of motion for soul; if flux does characterize soul, then the soul, for the purposes at hand, is ontologically no different than the phenomena. It turns out, though, that the soul while in one respect is like the flux as being in motion, in another, insofar as it is rational, is in part at rest, and so is in this respect like its objects, the Ideas. This (seeming) paradox also appears in the *Timaeus* where it is said that the soul is in part made out of the Ideas and in part made out of the phenomenal flux (35a); it is resolved not by claiming with Cherniss that the soul here is autokinetic and the exclusive instantiation of the Idea of motion,[12] but rather simply by noting that the motion of rational soul is the uniform motion of rotation, which is at rest with respect to its center. The phrase τὸ κατὰ ταὐτὰ καὶ ὡσαύτως καὶ περὶ τὸ αὐτό does not look backward to 248a, but rather describes the rational soul's uniform motion, which as such distinguishes the soul both from the Ideas and the phenomenal flux. The phrase τὸ κατὰ ταὐτὰ καὶ ὡσαύτως . . . πρὸς τὰ αὐτά describes the uniform rotary motion of rational soul at *Laws* 898a8–9,[13] and κατὰ ταὐτά is an essential attribute of the

---

[10] Cherniss (7) n40.

[11] A weakness of this whole approach is that, however read, 249b8 ff. is a list of dogmatic assertions. As such, these lines do little to counter the first argument, if in fact the first argument has any force at all *as an argument*. Even if 249b8 ff. reads as a recantation—and it does not—the recantation would have to offer reasons why the earlier argument was wrong.

[12] Cherniss (7) 311–12.

[13] On this passage, see Lee (4).

motion of the rational World-Soul in the *Timaeus* (36c2). Without such motion souls would not be rational. Our rationality is expressed in rotary motions that imitate the rotary motions of the heavens:

> God invented and gave us vision in order that we might observe the circuits of intelligence in the heaven and profit by them for the revolutions of our own thought, which are akin to them, ... and that by learning to know them and acquiring the power to compute them rightly according to nature, we might reproduce the perfectly unerring revolutions of the god and reduce to settled order the wandering motions in ourselves (*Timaeus* 47b6–c4, Cornford)

Our own rationality then guarantees that there will be motion which is not flux (*Timaeus* 47b-d) and so also guarantees that motion is part of complete reality. This sought-for conclusion is achieved even if not all reason dwells in soul.

We can then let stand the *prima facie* evidence that Plato did not consider God to be a soul. One could try to resort to an argument from silence, claiming that in the *Timaeus, Statesman,* and *Philebus,* Plato in fact did believe that God was a soul, but failed to mention it in these texts since there was no good reason for him to do so. Such a move, though, is considerably weakened if one cannot produce a clear instance somewhere in the corpus where the single, rational Demiurge is said to be a soul. In the *Sophist* itself, near the end, we are told that there is a demiurgic God with divine knowledge (265c), but again here there is nothing to lead us to believe that this demiurgic God is a soul and yet he is said to make all the ensouled creatures, plants and animals. It seems then that Plato wishes to claim that that which produces soul is not itself a soul and that which is the source of structure and composition is not itself a composed entity, as is soul. These would be good reasons for Plato not to hold that God is a soul.

I think Hackforth is on the right track in asserting Plato's general motivation for holding that the Demiurge is not a soul:

> To identify him [the Demiurge] with ψυχή would be to deny his transcendence or externality, since ψυχή is a principle operative only in the realm of κίνησις and γένεσις: and thereby to deny his perfection, since perfection does not and cannot belong to κίνησις and γένεσις.[14]

I would formulate this motivation as follows: insofar as Plato considers the Demiurge to be a necessary existent whose essence is rationality, then it behooves him not to have that rationality inhere in soul, which has the potentiality for being irrational. Comparing the Demiurge to the World-Soul here is illustrative of the point. The rationality of the World-Soul as a qualification of soul is clearly subject to the buffetings and dis-

---

[14] Hackforth (1) 447.

ruptive incursions of the bodily, which cause the soul to become non-rational (*Statesman* 273a–d, and for rational human souls, see *Timaeus* 43a–c, 47b–c). By contrast, though the Demiurge in his craftings of the World-Body is not able to triumph completely over the inherent cussedness of the corporeal, nonetheless he, unlike the World-Soul, is not corrupted by the corporeal. The Demiurge's essential rationality and capacity to act as an initiator of order and rationality in others is assured, then, only if he is not a soul. If the Demiurge is to serve as a crafting agent who constantly works to bring that which falls away from a paradigmatic Form into accord with it, then he can be assimilated neither to a contingent soul nor (along Neoplatonic lines) to the Forms themselves.

ELEVEN

The Platonic Theodicy: *Laws* X, 899–905*

This chapter offers some general reflections on the nature of *Laws* X as a theodicy. I wish to suggest that Plato in the *Laws* X advances a solution to the problem of evil that is quite different from that in the *Timaeus.*

Cherniss writes of his own mode of reconciling the *Timaeus* and *Laws* X "that this account fails to solve the 'problem of evil', and so it does if evil is a problem to be solved only by demonstration of its non-existence or by moral justification of it as a necessary condition for the existence of good. Either of these 'solutions' would have appeared to Plato to be an immoral falsification of the data of experience. Evil, like other phenomena, he regarded as something to be explained, not to be explained away."[1] Insofar as the erratic motions of the *Timaeus* are a permanently disruptive factor in the created world, operating over, against, and in spite of demiurgic reason, Cherniss' appraisal of Plato's view of the problem of evil is correct. For on this account the evil represented by the erratic motions of the corporeal cannot be explained away either as non-existent or as a necessary condition for the existence of good.[2] This is not so, however, in the case of the *Laws* X, where in the little discussed passage 899d–905d, I suggest, Plato verges on explaining away evil.

Now, Plato does not claim that evil does not exist. There are and always will be evil souls (903d, 904c–e). But he does seem to suggest A) that evil is an appearance, an appearance not in the sense of being illusory (like a mirage or after-image) or in the sense of being a phenomenal appearance *as opposed to* a supersensible reality (like a Platonic Idea), but in the sense of being a property which may only characterize parts of a

---

* Reprinted from *Mind* 87 (1978) by prior contractual arrangement. Originally titled "Plato's Final Thoughts on Evil."

[1] Cherniss (6) 259.

[2] Chemiss and I are here using "necessary condition for" in a non-technical, quasi-causal, quasi-instrumental sense. In the strict extensional sense of "necessary condition for" the evil represented by erratic motions of the corporeal is indeed a necessary condition for moral goodness on Cherniss' account of the *Timaeus,* since on Cherniss' interpretation wherever morally good acts are, there too are erratic motions—set up as inadvertent but inevitable consequences of goods acts. Cherniss (6) 258.

whole but never a whole, whether its parts are good or bad or both. From the perspective of the whole of things there is no evil (*Laws* 904b).

Further, though Plato in the *Laws* X does not treat evil as a necessary condition for the existence of good such that without the existence of evil goodness could in no way exist, he does strongly suggest B) that evil contributes to the completeness of the whole of things, which in its completeness is good (905b–c).

If A) and B) are true, Plato's views in the *Laws* X lie somewhere between Idealist and Thomist solutions or dissolutions of the problem of evil.[3] I will now defend claims A) and B).

The *Laws*, like the *Timaeus*, interprets divinity as a demiurge (902e5). This Demiurge is called a king (904a6, as in the *Philebus*, 28c7) and is also called a guardian (907a2, 7). Unlike the Demiurge of the *Timaeus*, however, the god of *Laws* X seems to be omnipotent. He is not omnipotent in the sense of being able to do just anything and everything, like making false analytically true statements; for instance, it is not possible for the Demiurge of the *Laws* to make it false that an event which has happened has happened. Rather he is omnipotent in that his materials as materials offer no resistance to his demiurgic activities. This can be established by inference from 904a–b, as Taylor does by citing this passage and claiming "clearly it is meant at least that there is nowhere in the universe any independent power which can cause this divine purpose to fail of its intent."[4] But the omnipotence of divinity can be established more directly by noticing that the question of whether god has the power to care for and control all things is explicitly raised at 901c and is answered emphatically in the affirmative at 902e7–903a1. The reader fresh from reading the mechanism for maintaining the erratic flux of phenomena in the *Timaeus* (52d–58c, especially 52d–53a and 58a–c) ought to find himself rubbing his eyes upon reading the claim that "it is more difficult to see and hear small things than great; but everyone finds it more easy to move, control (κρατεῖν), and care for things small and few than their opposites" (*Laws* 902c). Even the casual reader of the discussion of Necessity in the *Timaeus* should be amazed at the scope and the powers assigned to the

---

[3] Only Taylor has noticed the significance of *Laws* 903–905. Taylor (2) 184, 193. However, Taylor uses these pages of the *Laws* as principles of interpretation for the *Timaeus* and so ends up giving a greatly distorted view of the nature and power of both Reason and Necessity in the *Timaeus*. Taylor (2) 192, 194–95, against which see Cornford (3) 325–28. Cherniss underinterprets *Laws* 903–905 by mentioning the passage only in his discussion of what he calls 'negative' evils, when the evils under consideration in the passage are in fact the positive vices of evil souls (see n2 above). Cherniss (6) nn14, 15.

[4] Taylor (2) 184; cf. 193.

various types of craftsmen at 902d–e: their materials offer no interference to their projects.[5]

Now given that the Demiurge of the *Laws* X is perfectly good (900d) and is not ignorant of evil goings-on (901d) and is also ready, willing, and able to eradicate evil (902e7–903a1) *and yet* does not eliminate evil from "small things" or "parts of the whole" (901b8, c1), we must assume that the Demiurge's action or rather inaction with regard to small evils is actually a choice for the best. And so it turns out.

It is claimed that "all things are ordered systematically by Him who cares for the World-all with a view to the preservation and excellence of the Whole, where of also each part, so far as it can, does and suffers what is proper to it" (903b4–7, Bury). This arrangement of parts and allocation to them of what is appropriate to each is specifically cashed out as the Demiurge's shifting of better souls to a superior place and worse souls to a worse place (903d, 904c–905b). This shifting, however, is not meant merely to indicate to the skeptical that there is divine justice, that those who succeed through evil deeds are ultimately punished (though this too is intended), such that it is merely the fruits of evil deeds that are illusory. Nor is the punishment of evil souls here meant primarily to serve a purgatorial or educational purpose. Rather, Plato is claiming 1) that evil is a property which the Demiurge permits to exist only in the parts of the whole but which he never allows to apply to the whole as a whole, and 2) that only the quality of the whole will be used as a criterion for whether its parts are successfully integrated into it (903c7–d1). Thus Plato claims that what turns out best for the whole is also best for all the parts (903d2), even though a particular part may on its own independently of its place in the whole be evil. Evil, then, though very real from the perspective of one who is doing or suffering it, is an appearance from the perspective of the whole. Thus one whose perspective is limited to only his own partial condition is said to be ignorant of his true condition (903c2, d1), and so too, though one is not deceived in one's observation of partial evils, one is mistaken if one feels anxiety over such partial evils (903d1). I take claim A), then, as being established. Plato in the *Laws* treats evil as an appearance in that it is a property (unlike goodness) which applies only to parts and not to wholes.

---

[5] The scope of divinity's powers includes corporeal objects (i.e., the objects of sensation, 901d) and souls, both good and bad (903d). Any hint of resistance to demiurgic activity suggested by the phrase εἰς δύναμιν . . . πάσχει (903b6–7) would seem to be obliterated by the corresponding expression μετατιθέναι . . . κατὰ τὸ πρέπον αὐτῶν ἕκαστον at 903d6–8, the thought of which is restated emphatically at 904b4–6.

Plato does not stop here, though. In other dialogues Plato conceives of wholes which are good as wholes and whose parts, also, are each good. Such is the condition of the state and of the soul in the *Republic*. Each part of the good state and of the good soul is good in the whole of which it is a part and in itself is good as well. Further, in the *Statesman,* the guardian statesmen, acting as craftsmen by bringing that which falls away from a standard into accord with that standard, take that which is in itself evil (timidity or temerity, for instance) and by bringing it into a demiurgically measured whole make it good both in itself (as courage or temperance, for instance) and in the complete state (306a ff.). Why then doesn't the guardian Demiurge of the *Laws* do the same, taking that which in itself is evil or degenerate and making from it a complete whole all the parts of which are good both in the whole and in themselves?

The answer seems to be that the Demiurge supposes the whole which he constructs of parts that are both good and bad is better than a whole of only good parts. The specific advantage to the whole of permitting evil souls to exist seems to be the preservation of something like the free-will of moral agents (τὴν αὑτῆς βούλησιν, 904d5). But in any case there are two passages where Plato suggests that evils do not just happen to be parts of a whole which is good, but actually have a role (unspecified) that contributes to the good of the whole. These passages establish point B). The first runs:

> To each of these parts [which include the evil doers of 899e–900a] down to the smallest fraction rulers of their actions and passions are appointed to bring about fulfillment even to the uttermost fraction; whereof thy portion, O perverse man, is one, and tends therefore always in its striving towards the All, tiny though it may be (903b7–c2, Bury).

This passage does not mean that the evil uttermost fractions are made good or complete in themselves and only then are grafted to the whole, but rather that they are appropriated directly as they are for the completion of the whole, as is made clear in the second passage:

> You imagine, therefore, that in their actions [those of the evil doers of 899e–900a], as in mirrors, you beheld the entire neglect of the gods, not knowing of their joint contribution[6] and how it contributes to the All (905b5–c1).

Here I suggest we are in country quite foreign to the *Timaeus.* It is not *qua* evil (erratic) that the accompanying causes of the *Timaeus* contribute to the work of the Demiurge. Though demiurgic Reason may appropriate the 'traces' of order that exist even in the pre-cosmos (53b2), it must persuade and organize Necessity before Necessity can be put to

[6] The antecedent of αὐτῶν (905b7) is minimally αὐτῶν ταῖς πάξεσιν (905b5–6), but may also include as well θεῶν (905b6).

work. The primary bodies as in chaotic flux serve no purpose (46e5–6), but as organized into, say, an eye capable of sight, the primary bodies are, then, capable of serving a demiurgically conceived purpose (47a).

But in the *Laws* X we see Plato at the head of a tradition a member of which would eventually write:

> Evil, as we say (usually without meaning it), is overruled and subserves. It is enlisted and it plays a part in a higher good end . . . . 'Heaven's design', if we may speak so, can realize itself as effectively in 'Catiline or Borgia' as in the scrupulous or innocent.[7]

If I am correct in these claims, then, the *Timaeus* is not a theodicy, if by "theodicy" we mean an explanation of the existence of evil *in light of* the existence of a god who is good. For in the *Timaeus* the existence of evil is not of the Demiurge's choosing; it exists in spite of the best demiurgic actions (as a necessary inadvertent consequence of the Demiurge's actions if you follow Cherniss' interpretation of the *Timaeus,* or as a permanent factor of a completely independent origin, if you follow my interpretation).[8] The *Laws* X, however, is a theodicy proper. It must and indeed does explain the existence of evil in light of the existence of divinity, since, unlike in the *Timaeus,* divinity is not, in the *Laws,* hampered by the flaws of its materials. In the *Laws* evil does not exist over and against and in spite of the Demiurge, but is adapted just as it is directly into his design.

---

[7] F. H. Bradley, *Appearance and Reality* (1893; London: Oxford University Press, 1969) 178; cf. 355.

[8] See chapter 6.

PART FOUR

RELATED ESSAYS ON PLATO'S METAPHYSICS

TWELVE

Family Resemblance, Platonism, Universals*

I

Platonic universals received sympathetic attention at the turn of the twentieth century in the early writings of Moore and Russell. But this interest quickly waned with the empiricist and nominalist movements of the 'twenties and 'thirties. In this process of declining interest, Wittgenstein's theory of family resemblance seemed to serve as both *coup de grâce* and post-mortem.

I propose, however, that family resemblance far from being an adequate refutation of Platonic universals can actually be accommodated within a Platonic theory properly conceived. But first for some caveats and qualifications.[1]

---

* Reprinted from the *Canadian Journal of Philosophy* 7 (1977) by permission of the editor.

[1] The reader should be warned of two opinions which I hold concerning Wittgenstein scholarship. First, I should like to signal agreement with a view expressed by Alice Ambrose to the effect that Wittgenstein's writings on family resemblance can, and should, be treated in isolation from the rest of his later writings. Ambrose 118. The reason for this view is that the theory of family resemblance offers at best a solution to the classical problem of universals rather than the desired dissolution of the problem, which is to be found in the rest of his late work. In solving the problem of universals, Wittgenstein is simply ringing changes on traditional nominalistic views, whereas the dissolving of the problem is a consequence of his theory of meaning. Ambrose 118. I intend to discuss the theory of family resemblance only along the traditional lines of the problem of universals and simply suspend judgment regarding the dissolution of the problem.

Second, there has been some confusion about what constitutes the very problem family resemblance is supposed to be solving. Renford Bambrough has hailed Wittgenstein's theory of family resemblance as having solved the problem of universals by resolving some and circumventing other differences between nominalism and realism. Bambrough, 109, 117–27. But the nominalist/realist distinction can be understood in at least two quite different senses. One regards the universal/particular distinction and one regards the essence/accident distinction. The traditional problem of universals is whether general terms or universals are understood as being equivalent in some sense to the particulars that make up their extensions (nominalism) or whether beyond the many particulars there is

What family resemblance actually succeeds in refuting is not Platonic universals but Aristotelian or empiricist, or, generally, abstractive or commutative, universals. An abstractive universal is a universal arrived at by induction from identical characteristics in numerically distinct individuals (thus, for instance, see Aristotle's *Metaphysics* V. 26, 1023b30–31). An abstractive universal is a common property and nothing else, a property that things have severally but in common—each of them has it and it's the same from one to the next. This conception of a universal has several consequences. First, abstractive universals are ontologically dependent on particulars. If particulars did not exist, such universals would not exist. Moreover, as an abstractive universal becomes more general in scope, it becomes more vacuous, for abstractive genera are predicable of species but not vice versa. But what immediately concerns us is that, as a common property, an abstractive universal is necessarily simple in relation to its particulars. That is, a common property must be predicable univocally of the particulars which possess it (see Aristotle's *Categories* 5, 3a33–b10).

If family resemblance is an accurate account of the usage of terms, then this account of universals must be rejected. The family resemblance theory holds that there need be no element common to multiple referents of the same term by which we could define the kind of the plurality of referents. Wittgenstein specifically denies "the idea of a general concept as

---

some one thing, in some sense of "thing," in relation to which particulars are what they are and by reference to which general terms have their meanings (realism). In setting forth the theory of family resemblance in the *Blue Book* as a cure for "our craving for generality" and for our "contemptuous attitude towards the particular case," it is clear that Wittgenstein is advancing his theory of family resemblance as a resolution of the traditional one-over-many problem and that family resemblance is relevant to this first sense of the realist/nominalist distinction. Wittgenstein (1) 17, 18. It is with this sense, too, that I will be concerned. Bambrough, however, bypasses this sense of the nominalist/realist distinction. Instead, he speaks of realism and nominalism in a second sense, one that regards essences. A realist with regard to essences believes in the objectivity of classificatory schemata, whereas a nominalist with regard to essences believes that classificatory schemata are matters of social convention. Cf. Bambrough, 117, 126. It seems to me that the doctrine of family resemblance *per se* has nothing to say on this issue, though Wittgenstein's general theory of meaning certainly does. Further, Herbert Schwyzer has shown that Bambrough's analysis, which assimilates "meaning" to "denotation," does not even succeed as an interpretation of Wittgenstein in general and that on the issue of classificatory objectivity, Bambrough finds realist touches which are not present in Wittgenstein's thought. Schwyzer 69–78. I will not then be concerned with Bambrough's approach to family resemblance and universals, but will be concerned with family resemblance only within the traditional context of the problem of universals.

being a common property of its particular instances."[2] Rather particulars form family groups whose members' characteristics are related in ways which criss-cross and overlap.[3] The family name, say, "game" or "tool," need not pick out a single character or set of characteristics which is the same from member to member. Thus the family name is not used univocally each time it is used to identify particulars. Neither, however, is the family name baldly equivocal each time it is used to identify particulars, as for example "date" is baldly equivocal in its usages in the phrases "date palm" and "calendar date." As neither equivocal nor univocal, a family name is complex in relation to the individuals it is used to identify.[4]

Wittgenstein would have us believe that since there is no universal as common property to articulate the way in which family members form a group, we must look only to the individual members for such an articulation: "How should we explain to someone what a game is? I imagine we should describe *games* to him, and we might add: 'This *and similar things* are called "games."'"[5] In this regard Wittgenstein is carrying on the tradition of nominalism with regard to universals. All statements about universals (family names) can be reduced to statements about particulars, particulars and similarities to other particulars. Wittgenstein, however, differs from other nominalists in that his particulars are not necessarily atomic. They have relations which criss-cross and overlap.

However, eliminating a universal as common property as the criterion for denominating family groups is not sufficient to eliminate all notions of universals. In particular it is not sufficient to eliminate Platonic universals understood not as common properties but as standards. A 'universal' as standard is that by comparison with which we may say that any particular is a particular of a certain kind. Given a standard we can denominate a grouping of individuals in relation to it. The Standard Meter Stick in Paris defines by reference to it the class of things that are a meter long. But standards define by reference not only "perfect" particulars, but also degenerate or variant particulars.[6] And it is for this reason

---

[2] Wittgenstein (1) 17–18.

[3] Wittgenstein (2) sections 66–67.

[4] My analysis here is similar to A. J. Ayer's. Ayer 10–12. Individuals exhibiting various gradations of a sliding scale, such as shades of a single color (Ayer's example) constitute a minimal case of individuals forming a family group. Bambrough's critique of Ayer, therefore, collapses. Bambrough 116.

[5] Wittgenstein (2) section 69, emphasis in the original.

[6] By a "perfect" instance I mean a particular that corresponds exactly to a standard. If I have an extremely elastic rubber band and proceed to stretch it from its relaxed length to the length of the room, at some one point in the stretching process it will correspond exactly in length to The Standard Meter Stick in Paris. At that instant, it is a perfect particular; at all others it is a degenerate particular.

that I wish to claim that universals understood as standards need not, though in some cases they may, denominate a class of instances which have a common property running through all them.

To this end I wish to suggest that the relation of a Platonic universal to a particular is analogous to the relation of a musical theme to variations on the theme. Given a musical theme, we can designate by reference to it that a certain string of notes is a variation on it. And we can do this a number of times for several strings of notes and thus demarcate several variations. A mere repetition of the theme would be a perfect particular of the theme, though since it is so defined in relation to the theme, it is not to be confused with the theme itself. However, there need not be anything in common to all the variations, even though they are all variations on a single theme. Witness the *Goldberg Variations,* the *Diabelli Variations.* And thus we have in the musical theme a universal that (1) demarcates groups of particulars, (2) gives the kind of each particular in that each is defined by reference to the universal, but that (3) does not necessarily entail that there is any common characteristic which the particulars severally but in common possess. Thus, we have taken up the elements of the theory of family resemblance into a Platonic theory of universals. A set of variations form "a complicated network of similarities overlapping and crisscrossing: sometimes overall similarities, sometimes similarities of detail."[7] But the variations still are definable with reference to one universal form as standard.

The universal as standard is not strictly univocal in relation to the variations on it. That is, the relation "being-a-variation-of" picks out something different for each variation. For what type of variation each variation is varies from variation to variation. One variation could be a harmonic variation on the theme; one a melodic variation; one a rhythmic variation. And yet the use of the theme to pick out individuals is not baldly equivocal from variation to variation, since all the variations can be identified by reference to the one theme of which they all are variations. Rather, like the names of families and unlike abstractive universals, 'universals' on a Platonic theory are complex with relation to particulars.

I would like briefly to show that the theory of universals which I have sketched is consistent in broad outline with the major epistemological and metaphysical commitments of the middle Platonic dialogues (Section II). And finally I wish to suggest that Wittgenstein misunderstood how standards operate (Section III).

---

7 Wittgenstein (2) section 66.

II

The process of seeing the relation of the type of Platonic universals I have sketched to their particular instantiations need not be unidirectional, i.e., from the universal to the particular. Given that a certain number of particulars do form a variation-group, one can come to know the standard that operates among them, even though there need not be any element common to all members of the group and even though their standard is not immediately set beside them. To do this for instance is the sort of function which police artists perform. From many various descriptions of an unapprehended suspect, which need not all have any one feature in common, the artist arrives at a composite sketch by reference to which, the suspect may be identified.

Moreover, even if all the descriptions did have some one feature in common, this alone would not constitute a significantly determinate detail, since, for instance, "red-haired" or "blue-eyed" would lead to many irrelevant arrests. Thus, this process of indirectly approaching the standard cannot be confused, with induction, for it does not proceed by abstracting characteristics which are identical from individual to individual. Rather, the process is that which Plato terms "collection," and which is effected not by abstraction, but by recollection (*Phaedrus* 249b–c with 265d).

That collection does not involve abstraction from exactly similar properties accounts for the fact that there are no strict criteria for the properties a phenomenal object must have to initiate the process of recollection (see *Phaedo* 73d–74e, where even dissimilarities and deficiencies may initiate the process). That collection does not proceed by abstraction also means that even a single particular may initiate the process. There need not be a plurality of similar particulars for the process to operate.

Along these lines, it needs to be noted that in our analogy to recollection the composite police sketch which would result from this process is not a picture, image, copy, representation, imitation, or replica of any one of the descriptions of the suspect. If the police artist was shown a gallery of Hapsburg portraits and he made a composite sketch from the portraits, one could use the sketch to show that any one of the portraits was a portrait of a Hapsburg, but one could also go to the Schönbrunn and use the sketch to identify other family members in a more satisfactory way than by having taken along any one of the portraits. Thus, the sketch is not to be identified by reference to one of the portraits as an image is identified with reference to an original. Rather, just the opposite is the case. Therefore, the retrieved standards as universals are of a higher logical type than their particular instantiations, but as standards they, unlike abstractive universals, are also logically prior to their instantiations. That is, to define a particular entails reference to a universal, but to define a universal does not entail reference to particulars. And this is so even though, as retrieved or recollected, standards are epistemologically posterior to their instances. However, that a variation-group is and is known as a group presupposes that a non-retrieved standard (1) is ontologically prior to the group and (2) was at one time known in its non-retrieved state.[8] Otherwise, to claim that certain particulars formed a group to begin with, before the retrieval process, would simply beg the question. Such are the commitments of Plato's doctrine of recollection.

III

Wittgenstein, of course, gives an analysis of standards in the *Investigations*.[9] But, this analysis, I think, is wrong or at least incomplete. If Wittgenstein had realized how standards indeed work he might have avoided asserting that his theory of family resemblance does away with universals *überhaupt*.

Wittgenstein writes: "This sample [the standard sepia patch] is an instrument of the language used in ascriptions of colour. In this language-game it [the standard] is not something that is represented, but is a means of representation."[10] I suggest, however, that the relation of a standard to its instances is similar to the relation of an original to copies or imitations of an original.[11] As such, standards are indeed as Witt-

---

[8] My illustrative analogy of the police artist's sketch of the Hapsburgs does not capture these qualifications—(1) and (2). To do so, the sketch would have to be equivalent to the Form of Hapsburg. But as a human artifact the sketch cannot be a Form. Further most critics would wish to deny that Plato supposed there to be Forms of individual persons or their families. There is no Form of John Paul Jones, no Form of the Joneses.

[9] Wittgenstein (2) section 50. Geach has preserved two statements on standards made in conversation by Wittgenstein. One is consistent with Plato's analysis of standards. It reads: "The bed in my bedroom is to the Bed [i.e., the Idea of Bed], not as a thing is to an attribute or characteristic, but rather as a pound weight or yard measure in a shop is to the standard pound or yard." Geach 267, cp. 269.

[10] Wittgenstein (2) section 50. Wittgenstein goes on to use this analysis of standards as an analogue to his analysis of the utterances of words: "And just this goes for an element in [a] language-game when we name it by uttering the word 'R': this gives this object a role in our language-game; it is now a *means* of representation." Wittgenstein (2) section 50, Wittgenstein's emphasis.

[11] The relation of original to copy is minimally a highly illustrative species of the relation of standards to their instances. Plato, of course, goes further and identifies his paradigmatic standards with originals and the instances of standards with copies or imitations of originals (e.g., *Timaeus* 48e5–6).

genstein claims instruments of representation, but only in that broadest of sense in which any *sine qua non* or material of an artist's craft could be called his instrument or tool. However, an artist's standard or model (the material *from which* he works) can be sharply distinguished both from his tools or instruments proper (the materials *with which* he works, say, a point and mallet) and from his materials proper (the materials on *which* he works, say, marble blocks). This tripartite distinction of *"from which," "with which,"* and *"on which"* holds also for cases of the usage of standards where we would not, at least in common parlance, also call the standards originals. Take, for example, a seamstress who wishes to cut a yard's length of cloth from a bolt by appealing to a yardstick as a standard. Now we may say loosely speaking, that she *uses* the yardstick to perform this task and so speak of the yardstick as a tool or instrument, but in this sense of "use" she also uses the bolt of material. The bolt and the yardstick are both objects which she manipulates in order to carry out her task, and without which she could not carry out her task. Strictly speaking though, the cloth is that on *which* she imposes measure, her shears are that *with which* she imposes the measure, and the yardstick is that *from which* she takes the measure. As such the yardstick is not strictly speaking a tool, but is that which gets instantiated or expressed in the cloth, just as an original is instantiated, expressed, or represented in that which is a copy, imitation, or image of it. When we appeal to a standard, not to make an instance of it (as in the seamstress example), but to *identify* an instance of it (as when we appeal to a yardstick to say how long something is which already has a determinate length), usually the only tools (in the strict sense of "tool") which we use are our hands, *with which,* say, we take a measurement. The yardstick is in this instance, also, that *from which* we take the measurement.

Wittgenstein's failure to distinguish "that *from* which" from "that *with* which" in his discussions of language-games which employ standards means that he is mistaken in saying that standards are strictly means of expression, but are not that which gets expressed or represented.[12]

In having overlooked the fact that standards are expressed or represented in their instances, Wittgenstein must also necessarily overlook the possibility that standards get expressed and represented in a variety of ways such that there are degenerate as well as perfect instances of a standard and so that a standard's instances may form variation-groups. Thus, Wittgenstein's analysis of standards saves Platonic universals from flinching at the stare of family resemblance.

---

[12] Wittgenstein's comparison of the usage of standards to the utterance of words (see note 10), therefore, collapses.

THIRTEEN

The Formation of the Cosmos in the *Statesman* Myth*

There is a classical debate in Platonic scholarship over whether, on the one hand, the temporal order of events in Plato's descriptions of the divine formation of the universe is to be read literally, and we are to take it that Plato believes that there was indeed a period in which the Demiurge had not intervened into chaos and had not created the World-Soul and other souls, or whether, on the other hand, these events are to be read analytically as being described solely for the sake of exposition, and we are to take it that chaos simply represents a constitutive factor within a world which always has been ordered and under the guidance of souls. This debate goes back to the first generation of Plato scholars within the Academy. Among modern critics, Cherniss has marshaled the arguments, many gleaned from Proclus, in favor of a non-literal reading of Plato's descriptions of the creation of the cosmos by the ordering of chaos[1] and Vlastos has marshaled the arguments in favor of a literal reading.[2]

The debate over literal versus non-literal interpretations of the creation act has largely been limited to the *Timaeus*. I wish to suggest that the largely neglected *Statesman* myth offers some support to literal interpretations of the creation act. As in the *Timaeus*, Plato in the *Statesman* myth at least *speaks* as though there was a definite creation act, so that the burden of proof rests on a critic who denies that Plato *means* that there was such an act.[3]

The Demiurge is said to endow the world with reason κατ' ἀρχάς, "in the beginning" (269d2; cf. *Timaeus* 48a5). Moreover, the language of 273b–d suggests a certain primeval state of chaos before there was any

---

* Reprinted from *Phoenix* 32 (1978) with permission of the editor.

[1] Cherniss (1) 421–31.

[2] Vlastos (3) 401–19. This article has been attacked and Cherniss' views defended by Leonardo Tarán, 372–407. Zeyl summarizes and obliquely accepts Vlastos' arguments for a literal reading of Timaeus' creation story. Zeyl (2) xx–xxv.

[3] Most non-literalist readings of the *Timaeus* seek to discharge the obligation by pointing out (alleged) doctrinal inconsistencies which are supposed to flow from the assumption of a creation act. See Vlastos (3) 405–406.

ordering of the universe by the Demiurge (especially b4–5, 6, c7–d1).[4] Cherniss wishes to claim that these indications of an initial creation act in the *Statesman,* as in the *Timaeus,* express "in the synthetic form of a cosmogony what is in fact an analysis of the constitutive factors of the universe."[5] Of the *Statesman* myth in particular, Cherniss writes: "In the *Politicus* . . . the precosmical disorder and the retrograde motion [are] mythical, i.e., factors of the actual phenomenal world isolated for the purpose of description."[6] But here Cherniss too quickly lumps together the initial pre-cosmic period of the *Statesman* myth with the many later retrograde cycles of the universe which are described in the myth. In the *Statesman* myth, though not in the *Timaeus,* the temporal isolation of separate cosmic periods, representing for non-literalists constitutive factors of the universe, is not poetically or mythically achieved by a unique creation act. Rather this separation is achieved by the periodic alternation of cosmic cycles. This alternation, however, need never have had a beginning. That the alternations are said to have a beginning is a fact over and above the bare existence of the alternations. The strong suggestion of an initial creation act in the *Statesman* myth, therefore, is not easily explained away by the same reasoning that is used to explain it away in the *Timaeus,* since in the *Statesman* myth the "synthetic form" which on a non-literal reading would express an analysis of the constitutive factors of the universe is sufficiently accounted for by the description of the alternations of cosmic cycles, and this description is independent of the description of the pre-cosmic period and the initial creation act. Therefore, though the initial creation act may not differ in kind from later demiurgic interventions in the *Statesman* myth, it is surprising that it should be mentioned at all, if it is not meant to be taken at face value. Taken as merely a literary device the description of the initial creation act in the *Statesman* is redundant and represents a literary excess uncharacteristic of the myth (see especially 269c5–270a8).

This redundancy has two unfortunate consequences for non-literalists. First, if the description of the initial creation act is not merely a literary device and yet for non-literalists has to be explained away as *meaning* just the opposite of what it *says,* then it looks very much as if Plato on a non-literalist account is intentionally misleading the reader.[7] Secondly, if the initial creation act in the *Statesman* is not to be reduced to a mere poetic device, then any attempt to explain away its literal sense by special pleading, though not impossible, would be inappropriate.

---

[4] See Skemp (2) 106.
[5] Cherniss (6) 254.
[6] Cherniss (6) 255n21.
[7] See Vlastos (3) 405.

It seems then that the presence in the *Statesman* *both* of the periodic alternation of cosmic cycles *and* of a pre-cosmic period with creation κατ' ἀρχάς makes a non-literal reading of the creation act *pro tanto* harder for the *Statesman* than for the *Timaeus*.[8]

[8] Johansen shows no awareness of the *Statesman* myth as he flirts with a non-literal reading of Timaeus' creation story: "Timaeus in his general talk of what is the case whenever god is absent might be suggesting that a temporal creation story is not a one-off event but possibly a recurrent event. . . . The temporal reading does not require that there be, in absolute terms, a first act of divine creation. The world can be eternal with a state of disorder temporally preceding each act of creation." Johansen 91. Such repeating cycles of creation and disorder are the very scenario entertained in the *Statesman* myth, which nonetheless also asserts an absolute first creation (269d2).

FOURTEEN

The Divided Line and the Doctrine of Recollection in Plato*

I. A PUZZLE

There is no explicit mention or even a hint of the doctrine of recollection in the *Republic.* Of pre-natal direct acquaintance with Forms and the regained familiarity with the memories of these acquaintances, effected by some sort of interactions with the phenomenal world, we hear not a word in the *Republic.* This absence of the doctrine is puzzling, for the central books of the *Republic* and especially the Divided Line passage which ends Book VI present an epistemology which is portrayed as being exhaustive of modes and objects of cognition.[1]

Our puzzle cannot be handsomely resolved with the possible hypothesis that Plato had abandoned the theory of recollection in the *Republic.* For the two-tiered ontology and epistemology which are the hallmarks of Platonism and which are integral parts of the doctrine of recollection are found in the central books of the *Republic* as well as in the *Phaedo* and *Symposium,* works written soon before the *Republic,* and in the *Phaedrus* and *Timaeus,* works written after the *Republic.* Further the doctrine of recollection itself occurs in dialogues which flank the *Republic*—the *Meno* and *Phaedo* before, the *Phaedrus* after. Now it is true that the doctrine of recollection does not occur in the dualist *Symposium,* but then the *Symposium* does not pretend to give an exhaustive account of epistemological modes and so one could retreat to an argument from silence with respect

---

* Reprinted from *Apeiron* 18 (1984) with permission of the managing editor.

[1] Terence Irwin has claimed to the contrary that the doctrine of recollection can be found in the *Republic,* that there is a "parallel in language and doctrine" between *Republic* 518b6–c2, d5–7 and *Meno* 84c10–d2, 85c4–7 and that the *Republic* passage states "a conclusion which Plato would be willing to draw from the argument of the *Meno.*" The *Meno* passages are, however, simply Socrates' disclaimers that he is not providing to the slave boy information that is sufficient for the boy to make his correct judgment. The *Republic* is making a different point in the passage which treats knowledge as a conversion of the soul. The point is that teaching is not a giving to the learner a *capacity, organ, skill, or methodology* by which he acquires information sufficient for making correct judgments. Neither the language nor doctrine of the passages are parallel and the *Republic* claim is not an extension or consequence of the *Meno*'s defensive claim, though the two passages are consistent with each other. Irwin (1) n5.

to this dialogue in a way one could not in the case of the *Republic* in order to explain the absence of the doctrine of recollection. So our puzzle remains.

I wish in this chapter to give a model of the mind which will reconcile the doctrine of recollection with the exhaustiveness of the Divided Line. It must be admitted that Plato in no one place gives us a comprehensive, positive analysis of the mind. All the models of the mind advanced in the *Theaetetus,* for instance, are all ultimately rejected. The model that I advance on Plato's behalf is collected together from a number of sources in the middle dialogues. The model relies heavily on consistency and coherency of exegetical data, but it is not constructed out of whole cloth. It is not merely an attempt to save Plato from himself. The model has the advantage of being able to explain major parts of Plato's epistemology and in particular is able to offer answers to several of the prickliest problems of the Divided Line: What are the objects of the third level of the Line, the objects of *diánoia*?[2] How is the mental transition from *diánoia* to *noûs* effected? What is dialectic? And how is it that mental capacities are for Plato merely aspects of the same entity differentiated only by their objects *and* "what they achieve" (477d1).

II. A MODEL OF MIND

Imagine the mind as a sort of container divided vertically into two compartments by a dark pane of glass. Imagine one chamber, call it the front chamber, with an aperture through which the mind apprehends the phenomenal world "with" the senses. View this front chamber as a seat of consciousness. Plato explicitly maintains the existence of such a unified seat of consciousness in the *Theaetetus* (184d–185d). The other cham-

[2] For the status of the question on this issue, see Smith 129–37. Critics have taken the following array of stances on the objects of the third level of the Line: 1) they are mathematical intermediaries; 2) they are some of the Forms; 3) they are all of the Forms viewed severally; 4) they are the same as the objects of the second level of the Line; 5) they are extensionally equivalent to the objects of the second level, only viewed as images rather than as material objects; 6) they are λόγοι or verbal images of Ideas; 7) there are no such objects. Burnyeat treats the objects of the third level of the Divided Line as though they were like Aristotle's mathematical objects from *Physics* II, 2, that is, as mathematical forms that exist in matter but which can be conceptually abstracted or separated from matter and motion. Extensionally they are the same objects as the objects of the second level of the Line only with their "mathematically relevant features" attended to by the mind's capacity for abstraction. Burnyeat thinks these abstractions provide an adequate basis for Platonic knowledge, leaving *sub silentio* the existence of the fourth level of the Line and its objects, the Forms, as pointless. Burnyeat 229.

ber, the rear chamber, is a sort of storage room. It stores the memories of the Ideas of which we have had direct acquaintance in pre-natal life. Plato is somewhat vague about the status of these memory traces. Some texts suggest that they are the same for all and are, if recollected, grasped with the same degree of clarity as the original acquaintances with Ideas (*Phaedo*); some texts suggest that not all people saw all the same Ideas and not all people saw with the same clarity the Ideas which they did see (*Phaedrus*). This tension within the doctrine of recollection is exacerbated by Plato's failure to inform us whether, when the Ideas are recollected, our souls are still incarnate *or* are fully discarnate (along some analogue to astral projection) *or* a bit of both. Plato does not answer the question "where are we when we think?" despite the spatial metaphors which permeate his metaphysics and epistemology. For my purposes, though, I shall assume that the memory traces of the Ideas in the back chamber have the same content as the Ideas themselves and have the same clarity when recollected as the original pre-natal vision of the Ideas. This assumption is consistent with the *Republic,* in which Plato advances with seeming indifference both mental and non-mental entities as the objects of the philosopher-king's knowledge: in a little noted passage at the start of Book VI Plato claims that the philosopher-king looks to paradigms *in his soul* as his model for the ideal city which he sets about creating (484c7–8) and this passage is equivalent in force and content to the passage which ends Book VII where it is said that the philosopher-king looks to the Form of the Good and takes it as his paradigm for ordering the ideal city (540a8–b1). So we have in the back of the mind everything there is to know, the Ideas, their qualities and relations.

Now imagine that these memories are, like the Ideas themselves, self-illuminating and self-projecting, that they project images of themselves through the dark glass partition and into the seat of consciousness. The seat of consciousness consists of a medium which receives images, and a "perceiver," the mind's eye, which may look either to the images on the medium or to the phenomena through the mediation of the senses. It is the images projected through the partition which I propose are the objects of the third level of the Divided Line. They are fuzzy mental conceptions, sort of rough templates, but they are not "abstractions" from the phenomena. These images are degenerate in content and clarity due to the nature of the partition.

At *Phaedrus* 250a5–b1, Plato specifically claims that we do possess such fuzzy concepts, which are present to the incarnate mind even prior to recollection and which are derived from pre-natal acquaintance with Forms rather than from sense perception: "Only a few remain whose memory is good enough; and they are startled when they see an image of what they saw up there. Then they are beside themselves, and their experience is beyond their comprehension because they cannot fully grasp what it is that they are seeing" (Nehamas). Much the same point is made in the *Republic* itself (505d11–e3) where it is claimed that every soul, even when it has no acquaintance with the Good, nevertheless divines that it is something, but is perplexed and can not adequately grasp what it is. Here it is claimed that some fuzzy concept or foreknowledge of the Good is present to the minds even of those whose thought is restricted to the phenomenal realm. The concept, though, is neither an abstraction from the phenomena nor an acquaintance with the Good itself.

It is the existence of these fuzzy concepts which makes possible the formation of hypotheses which neither have the phenomena nor the Ideas as their objects, and yet which in accordance with the schematism of the Divided Line (especially 534a), must have objects. It is these innate concepts which, though vague and un-grounded, indeed "unknown" (533c), when 'perceived' by the seat of consciousness, account for the possibility of the special theoretical sciences of Book VII (mathematics, geometry, astronomy, etc.). But the scope of the images also can account for hypothesis formation in other disciplines, such as physics, ethics, and aesthetics.

That Plato conceives of some faculty of the mind as consisting of a combined image-receiver and image-perceiver, a faculty which is different from both reason and opinion, is clear from the *Timaeus* (52b1–c1, on 'knowing' space). Here the faculty is compared to dreaming, as is the faculty of *diánoia* in the *Republic* VII (533b7–c3), where the scientist who operates at the third level of the Line is said to be dreaming about being. I take it that the dream analogy in Book VII is different than in Book V, where Plato seems to be making a point about the possibility of metaphysical verification (476c). Rather the dream analogue in Book VII, as elsewhere in the dialogue, is meant as a positive analogue to a mode of intellectual access which is akin to reason (*Republic* IX, 572a–b; cf. *Timaeus* 71b–e, on the liver as a receiver and reflector of dream images).

There are four positive comparisons to be made between *diánoia* and dreaming. First, the contents of dreams are images that are projections received from elsewhere into a medium; they are not the spontaneous, *ex nihilo* products of what we would call fantasy (see *Timaeus* 71b). And yet, second, dream images may be manipulated and creatively re-arranged: dreaming is, in part, an activity and is not just a passive reception of images. Third, the objects of dreaming—unlike the objects of reason, and unlike even the objects of opinion, viewed as images of Ideas—are mind-dependent; they are mental entities. Fourth, dream-objects, like the objects of *diánoia,* are apprehended on some analogue with seeing. It is important here to remember that for the Greeks one always *sees* a dream; one does not *have* a dream.[3] A dream is not a quality that attaches to the

---

[3] See Dodds 105.

mind nor is it a mere activity of the mind. Rather it is something perceived by the mind's eye. In Plato's analogy of *diánoia* to dreaming, the images projected on the medium in the seat of consciousness are the correlate of dream-objects.

Further, the images which are on the third level of the Line are clearly intended to be, in some sense, relevantly similar to the images on the first level of the Line. The images on this level—shadows and reflections on water or in mirror images (509e–510a)—are all what may be called non-substantial images, that is, they are images which unlike statues, photographs, and painting, but like television images, are dependent for their existence both on the persistence of their originals and on the existence of a medium over which they may flicker, but which does not contribute to their permanence, as would a material component.[4] The first condition—persistence of the original—is fulfilled on the third level of the Line by the permanence of the pre-natally induced Ideal memory-traces held in the rear compartment of the mind. The second condition—the existence of a medium—I suggest, is fulfilled by viewing the objects of the third level of the Line as images reflected in the seat of consciousness.

III. OPINION AND THE SECOND LEVEL OF THE LINE

This model is able to explain Plato's not well understood view of what opinion is, the cognitive mode of the second level of the line. Now, Plato does not tell us much about the nature of opinion, but when he does he uses rather careful formulations. The usual formulation is that the opinable or sensible, is apprehended (παραλαμβάνω) *by* opinion (instrumental dative) *with the aid of* (μετά) non-discursive (ἀλογόν) sensation (*Timaeus* 28a1–3). Now, it is true that opinion has the sensible as its object or *referent* (ἐφ' οἷς, *Republic* 534a5–6) and is *about* the sensible realm (περὶ τὸ αἰσθητόν, *Timaeus* 37b6). But Plato does not say, contrary to our usual empiricist sorts of intuitions, that the content of our opinions is *from* the sensible. Rather Plato says that the sensible is apprehended *by* opinion, *not by* sensation, though it is apprehended with the aid of sensation. Normally we would suppose that we 'know' the sensible by the senses. But this is not so for Plato. The point, I suggest, that Plato is making by his curious formulation "with the aid of sensation" is that our senses do not provide us with enough information to be able to make true judgments about the sensible. The extra information that is needed is provided by the Ideal, albeit vague, memory-images in the seat of consciousness. Judgments of number, relation, and moral and aesthetic value require in-

[4] On Plato and non-substantial images, see Lee (1).

formation about numbers, relations, and qualities, from the mind, even for the simple purpose of identifying instances of them in the phenomenal realm, since for Plato instances of numbers, relations and ethical and aesthetic qualities in the make-up of the phenomenal realm are technically non-sensible.[5] Rather they supervene upon sensibles (colors, sounds, and the like, and possibly natural kinds) but they neither are proper, direct objects of sensation, nor are they non-proper but incidentally direct objects of sensation (as they are for Aristotle, *De Anima* III, 1). That we have vague notions of these concepts independently of sense experience allows us to make true and useful judgments about the world without resorting either to holding empiricist contentions that true beliefs are justified by the senses or to making the possession of Platonic knowledge—direct acquaintance with Forms—a precondition for having Platonic opinion.

Opinion, for Plato, then is neither seeing nor is it *like* seeing, in the way knowledge is for Plato *like* seeing, only a seeing with the mind's eye. Opinion, rather, is a *manipulation* of both innate ideas and sense impressions. So I suggest that when Plato says that opinion and knowledge differ 1) with respect to their objects *and* 2) with respect to 'what they achieve', this second condition is not vacuous, as it is for many critics, who take the second condition simply as the actualizing of the first condition, which for them is our potential for direct acquaintance with different orders of objects. The shift from potentiality to actuality is for most critics the same sort of shift in both the mode of knowing and the mode of opining. On my model the activities themselves of knowing and opining are different in kind, for opinion is not a form of direct acquaintance.

IV. THE DOWNWARD PATH AND THE THIRD LEVEL OF THE LINE

Given the model of mind which I have sketched, the "downward path" and the method of hypothesis can be handsomely explained. The geometer or ethical theorist can come up with hypotheses from the images of Ideas in the seat of consciousness. The images provide the content for hypotheses. That these images are derived from the Ideas keeps the hypotheses from being arbitrary. Plato is not suggesting, I think, that

[5] I am speaking of justice, beauty, relations, and numbers from the "argument from opposites" in the *Republic* (V, 479a–c, cp. VII, 523a–525a), justice and temperance and the like from the *Phaedrus* (250b–d), the common terms from the *Theaetetus* (185b–186a), and the great kinds (being, sameness, difference, motion, rest) from the *Sophist* in contrast to direct objects of sensation (sights, sounds, smells) and to natural kinds (tree, duck, planet, water).

the method of hypothesis is merely a matter of random guessing, followed merely by checks for consistency. Nor are the hypotheses related in a question-begging sort of way to other hypotheses, as is the case when a science is operating in a restrictive Kuhnian paradigm, wherein the range of likely hypotheses is largely predetermined by already established laws and methods. Further the model explains why Plato can claim that the hypotheses of geometry, say, are so widely held throughout the geometer's discipline (510c). The image content of the mind will be more or less the same for all, since the Ideal-memories are more or less the same for all.

Usually the method of hypothesis in the *Republic* is taken to be much like the process of Greek mathematical analysis, whereby a mathematician picks an arbitrary hypothesis, and if the hypothesis does not produce contradictions either among its propositional parts or in conjunction with other hypotheses, it is tentatively accepted as true. The downward path is, on this view, a natural deduction system applied to a hypothesis; as long as results of the deduction are self-consistent and consistent with other hypotheses and deductions from them, the hypothesis is retained.

In such a system, diagrams, and sensuous representations in general, serve only as do the diagrams in Euclidean geometry handbooks. The geometer uses them only as aids to a weak memory, just as a chess player with a weak memory uses chess-men and their positionings as tokens for what in fact are mental operations. Sensuous representations on this traditional model do not add anything to the content of the hypotheses.

If two hypotheses are found to be consistent, a scientist looks for a more general hypothesis which will encompass and explain collectively what the two hypotheses severally explained. Such a system of hypotheses neither arrives at a ground which is non-hypothetical nor achieves a degree of integration among hypotheses that is anything more than mere consistency.

The model which I have suggested can explain much more than mere consistency. Indeed it explains operations more like those of actual scientists and rational ethicists. We know that Plato supposes the Ideas to exist *all* in a rational plan (τεταγμένα . . . κόσμῳ πάντα καὶ κατὰ λόγον ἔχοντα, 500c2–4). He asserts this prior to and independently of whatever sort of integration the Idea of the good affords the other Ideas (508b, 509b). If this arrangement of the Ideas is imaged, however dimly and fragmentedly, in the seat of consciousness, then the scientist will be able to come up with hypotheses that dovetail with each other, overlap with each other, are internally related to each other, and form parts of systems with each other. The scientist on this model is quite capable of arranging and systematizing hypotheses, adjusting some, rejecting some, and positing new ones, based not merely on consistency, but based on a coherent plan which is partially revealed to him by the image content of the seat of consciousness.

The Platonic scientist operates with hypotheses a bit like an archeologist who is reconstructing an urn from a group of related potsherds, which may, though, be largely incomplete and which may contain foreign bits. He aims for a coherent plan which is suggested by the shards themselves. But, the coherency he achieves may not be the coherency of the original; that the hypothesis-shards do cohere in some manner at the end of the scientist's fashionings is not a guarantee of their validity. They remain, even when arranged, unknown and lacking a non-hypothetical ground. Plato is not committed to a naive coherency model of truth, wherein statements are true to the extent that they form a coherent story. Nevertheless, the hypothesis maker's manipulations accomplish much more than mere consistency; the scientist is not simply tied to *reductio* arguments and the Socratic elenchus, as these techniques are ably exemplified in Book I of the *Republic.* Such procedures could never have achieved the results of Books II–IV, which, in their sifting and sorting, integrating and adjusting common opinion, guesses, hunches, and folk wisdom, culminate in the four highly integrated, yet hypothetical, definitions of the virtues, which end Book IV.

Irwin contends that the sort of manipulations which I have suggested are involved in the method of hypothesis are really all that Plato intends to be accomplished by recollection: "On this view [of the theory of recollection] all that can be expected, and all that is needed to produce the most coherent system of mutually justifying beliefs about moral *knowledge,* is the mutual adjustment of beliefs about virtue and about the good."[6]

Plato is quite clear though that the integrated definitions of the virtues at the end of Book IV are tentative and hypothetical (433b, 434d, 435c–d, 443c, 543d–544a). Indeed, in a line that adumbrates the central books, the

---

[6] Irwin (1) 771, my emphasis. Irwin and Fine suppose that the only difference between cognition on the Third Level of the Line and that on the Fourth is that propositions that were viewed as a merely consistent with each other on the Third Level are on the Fourth viewed as mutually supporting each other—and that is what knowledge is, having in mind mutually supporting propositions. Fine calls this Platonic coherentism. But such coherentism doesn't require any special metaphysical entity, certainly no eternal, non-material entities of the sort Platonic Forms seem to be. Even nominalists who think that meanings are mere social conventions are willing to say that some, indeed probably most, social conventions, and so meanings, mutually support each other. Irwin and Fine's interpretation of the Divided Line simply turns Forms into so much excess ontological baggage. Plato may as well be a sophist on their interpretation. Irwin (3) 169, Fine (2) 14–15, 25.

definitions are said to have the status of "some sort of image of justice" (*Republic* IV, 443c4). They are not an expression of ethical knowledge, however well-integrated a set of beliefs they may be. The method of what Irwin calls "recollection," a method which he wishes to divorce from 1) Forms as external certain standards and 2) direct acquaintance with such Forms, in fact is, at best, only what Plato describes as the method of hypothesis, which itself for its success is dependent on the very premises (1 & 2) which Irwin wishes to reject.[7]

V. RECOLLECTION AS THE SHIFT FROM THE THIRD TO THE FOURTH LEVELS THE LINE

We are not told a thing in the *Republic* about how the transition in modes of cognition from the third level to the fourth level of the Line occurs. I suggest that recollection fills this gap. Something like the following occurs to effect the transition. The method of hypothesis is not sufficient to effect the transition (511a), for this process "can not reach beyond the hypotheses." But (δέ, 511a6), insofar as the method of hypothesis uses sensible representations, it is able, I suggest, to effect the transition. Plato is quite explicit that the phenomena as images of Ideas bear information which is useful to the scientist who is trying to get beyond hypotheses (510d5–8).

I think it likely that the phrase "visible εἴδεσι" (d5, form, figure) does not merely mean "figures" like sketches in a geometry book, but is Plato's technical use for "immanent characteristic" or "immanent form" (*Phaedo* 103b5, 8, e5; cf. *Philebus* 64a1–2).[8] The use to which these "forms" is put is not much elaborated; it is clear though that they are not mere physical tokens serving simply as aids to memory and conveying no information. For the aspect of the phenomena which the Divided Line stresses is the relation of the 'visible forms' to the Ideas themselves—the models of which the visible figures are likenesses (510d7)—and not the relation of the visible figures either to opinion formation or to hypotheses. Indeed we are told that when the scientist speaks of the visible images, he is not thinking about them but of (περί) the Ideas and is making λόγοι for the sake of (ἕνεκα) (knowing?) the Ideas.

Plato tells us more of the positive role which the phenomena play in acquiring reason later in the discussion of astronomy. We are told that

---

[7] Irwin (1) 772.

[8] Alexander Nehamas has argued, correctly to my mind, that the immanent characteristics of the *Phaedo* accurately image all the content of the Ideas of which they are instances, that they are not somehow necessarily degenerate simply by being phenomenal entities. Nehamas 105–17, especially 116.

the phenomenal realm, or at least the ordered part of it ("the embroidery of the sky"), serves as paradigms (παραδείγματα) in the study of Ideas (529d7–8).[9] Plato elaborates the import of these "sensible and material" paradigms (530b3) by saying that they are useful "just as if one comes upon plans (διαγράμμασιν) most carefully drawn and excellently worked out by the sculptor Daedalus, or some other craftsman or artist" (529d8–e3). Plato seems to be viewing these sensible paradigms something like blueprints. And indeed one of the standard senses of παράδειγμα is "an architect's plan" (*LSJ*, s.v.). Now a blueprint is not a house, though it can be used loosely as a standard for identifying houses. It is this aspect of the carefully drawn plans of Daedalus which explains why they are called "paradigms." More importantly though for the immediate purpose of the passage, which is to suggest how the phenomena aid in coming to gain a familiarity with the Ideas, the relevant feature of blueprints is that they are highly informative about houses, clearly and distinctly reflecting salient features of them. Viewing a blueprint is extremely useful in coming to understand what it is to be a house. This is why, for example, blueprints or floor plans so frequently appear in advertisements for atypical housing (condominiums, lofts, etc.)—to convince the buyer that he in fact is purchasing a house. Plato seems to be claiming then that the phenomena, or at least some phenomena, clearly and distinctly reflect salient features of the Forms of which they are instances and so offer clues to understanding what it is to be any given type of thing, to apprehending the Platonic Form.

Speech and discourse may have the same effect as visible images in serving in this way as blueprints. In the *Statesman* both speech and visual images are instrumental in indicating Ideas (285e–286a); indeed some Forms can only be gotten to with the aid of speech, for they do not have visible images. And in the *Meno* and *Phaedo* both speech and objects can spark the process of recollection.

Turning to the model of mind which I sketched earlier, I suggest that a joint combination of the information gleaned from the phenomena, viewed as a sort of blueprint, along with the information which the mind possesses as the result of holding images of the Ideal memory traces, has the effect of sparking recollection by knocking a hole in the semi-transparent partition which separates the Ideal memory traces from the seat of consciousness. Neither type of information alone is sufficient to this

---

[9] For sensible paradigms, see also *Timaeus* 29a2. In both the *Republic* passage and the *Timaeus*, it is the measures or standards of time (days, months, years, etc.) which Plato has in mind as central examples of what he means when he speaks of sensible paradigms (*Republic* 530a7–8, *Timaeus* 37a–38e). See chapter 2. By sensible παραδείγματα, then, Plato means some sort of "exemplars" or "standards" rather than "examples" or "parallel cases".

purpose. With a hole in the partition, the seat of consciousness has an unobstructed vision of at least some of the Ideal memory traces. If the traces are viewed as having the same status as the Ideas themselves, this direct acquaintance of them is Platonic knowledge. It is a kind of seeing, and it is non-discursive, even though *lógos* formation and hypothesis manipulation were instrumental in effecting recollection. Note especially *Timaeus* 28a1, where *lógos* stands to knowledge as sensation stands to opinion: it is that *with the aid of* which reason comes about. I take the seeing and touching metaphors which Plato uses throughout the central books of the *Republic* as serious metaphors, that is, as suggesting some positive analogue to reason and not as merely sloppy writing or as appeals to the sloppy customs of ordinary discourse (contra, say, Gosling) (see especially *Republic* VI, 490a–b).[10]

Knowledge and dialectic are not for Plato a form of mental processing, in the way I have suggested that opinion is a sort of processing of received information. In particular, I suggest, dialectic or knowing is not like conceptual analysis, whereby we feel that we understand a concept once we have mentally processed it into its constituent parts and their relations. Now this interpretation of mine at first may seem implausible if one believes that when Plato describes dialectic as "proceeding by means of Forms (dative) through Forms (διά + gen.: spatial, causal, manner?) in Forms and ending in Forms (εἰς)" (511c1–2), he means either the application of natural deduction rules to absolutely certain premises *or* means a process of analysis. This interpretation may additionally seem implausible if one supposes, as I indeed do, that when Plato speaks of collection and division of Forms, he means that collection and division have the Forms as their objects, rather than that, as Cherniss holds, collection and division are merely heuristic devices which assist in the process of recollection but which do not map the structure of Ideas.[11] I wish to suggest that both the view that dialectic is a movement or progression (ἰοῦσα, 510b7; καταβαίνῃ, 511b8) and the view that collection and division are about Forms can be accommodated within the direct acquaintance model of knowing.

One simply needs to imagine dialectic on an analogue with the way in which we view paintings. We do not see the whole of a painting and all its detail all at once. A good painting naturally leads the eye around the canvas. A series of shapes or color gradations will have the effect of causing us to pay attention to different types of structures and different groupings within the painting. Our eye moves back and forth from detail to composite and our attention constantly shifts from one type of structure to another. Such, I suggest, is the seeing of the mind's eye for Plato.

---

[10] Gosling (1) 120–23.

[11] Cherniss (2) 55.

The field of Ideas, as the result of their existing all in a plan, leads our attention from one Idea to another and causes us to see them as simples (division) and as parts of complexes and structures (collection). Dialectic is the tracing of relations between Forms; but it is a following of the relations suggested by the Ideas themselves. It is not a mental construction of the relations or a processing of information gleaned from acquaintance. Rather in the tracing process which makes up dialectic, the Ideas serve both as sign-posts and termini.

FIFTEEN

The Number Theory in Plato's *Republic* VII and *Philebus**

In this chapter I argue that the number theory that Plato advances in the late *Philebus* (56d–e) is the same theory that he articulates in Book VII of the *Republic* (526a). I suggest that both passages advance the Ideas or Forms of numbers as the objects of mathematical study and that neither passage entails a doctrine of mathematical abstractions from phenomenal particulars (in the manner of Aristotle's *Physics* II.2) or a doctrine of mathematical intermediaries. In the *Philebus* the objects of philosophical mathematics are formally introduced as paradigms for what is to count as what really is (59d4–5) and as the objects of reason (59d1). Therefore, if my interpretation of these passages is plausible, it should give pause to those critics who claim that in the *Philebus* Plato abandons, or at least is no longer obviously committed to, the two-worlds ontology of the central books of the *Republic*.[1]

In the *Philebus* Plato distinguishes the objects of the philosopher's mathematics from the mathematical objects of the masses (τῶν πολλῶν):

> Arithmeticians of the first sort [the populace] reckon unequal units, for instance, two armies and two oxen and two very small or even the very largest of all things; whereas arithmeticians of the other sort refuse to agree with them unless it is posited that each unit of the myriad [units] differs not at all from each and every other unit. (56d9–e3)

This passage parallels the *Republic* passage:

> —Then what do you think they would answer if someone asked them [the mathematical experts]: "You strange people, what kind of numbers are you talking about in which the one is as you assert it is, each being equal to all others without the slightest difference, and each containing no parts?"

---

* Reprinted from *Isis* 72 (1981) by permission of the University of Chicago Press.

[1] For example, Shiner and Gosling (3). Both authors think there is nothing in the *Philebus* that commits Plato to a belief in the theory of Ideas as formulated in the middle dialogues. Neither, however, gives an interpretation to *Philebus* 56d–e. See Shiner, 78, Gosling (3) 127. J. M. Moravcsik has recently given an interpretation of the early part of the *Philebus* (through Steph. 26) in which he contends that "the Theory of Forms turns out to be a presupposition of the views expounded here." Moravcsik 100–101.

—I think they would answer that they are talking about things which can only be grasped by thinking and which cannot be dealt with in any other way. (526a1–7)

I think the unjaundiced eye would admit that the two passages seem to articulate the same doctrine, whatever the doctrine is. Though the *Philebus* passage is telegraphically formulated, the content of each passage is so specific that the similar language of each can only be stating the same doctrine. The shorter *Philebus* passage is a blind reference to the longer *Republic* passage.[2]

I suggest that the doctrine contained in both passages is one half of the doctrine that Aristotle expounds as the Platonist doctrine of the inassociable or inaddible numbers (*Metaphysics* 1080a5 ff).[3] The doctrine briefly is that some numbers are Ideas which do not consist of units and which have no merely countable parts, that is, parts having no other content than their countability: oddness and primeness are each *one* aspect of three and therefore are countable parts of three; each is not, however, a merely countable part of three. These numbers neither are parts of other numbers nor have other numbers as their parts. Such numbers since they possess no units cannot be added or be subject to any other arithmetical calculations. Because these numbers stand entirely outside each other, their numerical content or differentiation is found in their differing positions in a definite serial order of priority and posteriority. The Ideas of numbers are the natural number line.

Taking this doctrine of the inassociable numbers and wedding it to the number theory of the *Phaedo* (101b–c, 102a–105b), we get the following relation between the Ideas of numbers and numbered objects. It is by reference to the ordinality of the Ideal realm that we are able to determine the cardinality of the phenomenal realm, the world of appearances. We can say that one grouping of phenomenal objects is larger in number than another by visual inspection, by mapping one set to the other without being able to say what number the number of each group is, just as we can tell by the senses that one body is taller or hotter than another without knowing what height or temperature either of them is. It is our ability to compare the cardinality of phenomenal groups without knowing what the number of each group is that leads Plato in the *Statesman*, somewhat to the reader's surprise, to rank number along with length, breadth, and thickness (B text; or swiftness, T text) as objects of the crafts that measure things in relation to their opposites (i.e., as being subject to

---

[2] The *Philebus* passage is a blind reference to the *Republic* VII passage in the way that at *Phaedo* 73a–b the mention of the use of a diagram in the processes of recollection is a blind reference to such use of a diagram at *Meno* 82b–c.

[3] See Cook Wilson, 247–60, Cherniss (1) 513–24, Annas (1) 17–18.

the more-and-less) rather than in relation to measures or standards (284e). We determine what number a phenomenal number is, not by comparing it to another phenomenal number, but by referring it to the standard or Idea of number of which it is a representation. And we do this not by mapping the phenomenal units in the numbered grouping to (alleged) units within the Idea of number of which it is an instance; rather we map the units of the phenomenal number, one each to each position on the number line starting with the Idea of one. Thus the number of fingers on my left hand is five, since this mapping process is completed at the fifth Idea of number. This is about all of the Platonic number theory that can be gleaned from the Platonic texts and from those of Aristotle's reports that are consistent with the Platonic texts.

The half of this number theory to be found in our two texts is that the Ideas of numbers have no merely numerical parts determining their numerical content and are not themselves numerically determining parts of each other; or in a word, they are inassociable or inaddible. The other half of the theory—that the Ideas of numbers form a determinate serial order, which does determine the numerical content of the Ideas of number—is apparently not to be found in the Platonic corpus. Aristotle, however, attests that the inassociability of numbers and the determinate serial order of numbers are necessary and sufficient conditions for each other.[4]

It should be clear enough that the numbers without parts at *Republic* 526a are Ideas of numbers, in that they are called objects of reason itself (525c3, cf. 523a), are said to belong to the realm of being in opposition to becoming (525c5–6; cf. 526a, 527a5, 529a2, 529b5), are non-sensible as opposed to numbers with visible and touchable bodies (525d7–8, cf. 529b5), and are called numbers themselves (525d6) and are therefore on the same footing as "the square itself, the diameter itself" (510d7–8), and justice itself (517e1–2). Further, therefore, they are at the fourth and highest level of the Divided Line (which marks off the intelligible world from the phenomenal world) and belong to "what I call the intelligible class" (513a3). Wedberg nonetheless tries to argue that *Republic* 525d–526a refers not to Ideal numbers, a notion he maintains "is not explicitly recognized in the *Republic*," but to mathematical numbers or mathematical intermediaries, which allegedly hover between the Ideas and the phenomena and which are supposed to be perfect non-sensible particulars of the Ideas of numbers, perfect instances of which allegedly cannot occur in the phenomenal realm. Wedberg's sole evidence for this assertion is that the reference of "calculations" at 525d1 is to the arithmetical prac-

---

[4] See Cherniss (1) 514.

tices of Plato's day, which supposedly entail mathematical numbers.[5] But in fact this occurrence of "calculations" is a reference back to 522c6–7 and 525a9, and so clearly awaits the passage under question (526a) to explain what the numbers indeed are that calculations studies. On this passage (526a) Wedberg simply shrugs his shoulders, saying that in the absence of the Ideal theory "Socrates is here unable to make his argument clear." But when Socrates does refer to popular arithmetic, he has in mind admirals and generals and the like counting "numbers with sensible bodies" like ships and troops, rather than non-sensible "ideal" units (522d–e).

The structure of the whole discussion of mathematics as a special theoretical science (VII, 522b–526b) presupposes that only two sorts of numbers are being discussed. The structure reprises the purely dualistic argument for Forms from the end of book V—479a–c, the "argument from opposites"—to show that mathematical claims cannot be adequately explained by reference to numbers with visible and touchable bodies and so the study of mathematics drives us to hypothesize Forms of numbers. The disjunctive argumentative structure collapses if there are intervening levels of numbers between the phenomena and the Forms.

Our passage, 526a, then, is discussing Ideas of numbers. Further, the qualities it ascribes to them are peculiar to the Ideas of numbers *qua* numbers, not *qua* Ideas: the passage is dealing not with the formal or metaphysical properties of the Ideas of numbers as Ideas, but with their particular content as differing from other Ideas.[6] Cherniss argues to the contrary that our passage is dealing with Ideal numbers only as Ideas and is not dealing with their special content, such that for Plato "the Ideal numbers were first conceived as incomparable because they are Ideas." For Cherniss no Ideas have parts; they are in every sense simple

---

[5] Wedberg 124. It is true that Glaucon says the numbers in question are grasped by *dianoēthēnai* (526a6), where one might have expected the prefixless *noeîn*; and so it might appear that Plato is suggesting that he has in mind objects of the third, not the fourth, level of the Divided Line. But as Plato admits at the end of the discussion of the special theoretical sciences, the language he uses of the Line is extremely liquid and not to be used or interpreted rigidly, especially when discussion of the four-tiered Line itself is not in immediate view (533d–534a). In any case, not four lines after the allusion to *diánoia*—with no intervening numerical content between—Socrates, perhaps as a gentle corrective to Glaucon's choice of verbs for thinking, uses *noēsis*, the mode of cognition of the fourth level of the Line, to describe the mode of cognition by which the numbers under discussion must be known (526b2).

[6] On this point it is important to notice that the reintroduction of the "argument from opposites" of Book V (479a–480a) into Book VII—an argument that applies to the whole range of Ideas and their instances (479a–b)—formally terminates at 525c6. The καὶ μήν of c8 introduces a new topic, unique to calculation.

and indivisible. Of our passage in particular he contends that "'each unit' here is the unity of each of the numbers which are accessible to intelligence alone and which, as indivisible units, are all 'equal' and quantitatively indifferent"; namely, it is as Ideas that the numbers under discussion are indivisible.[7] However, the phrase "in which the one is as you assert" (526a2–3) is fairly clearly a restrictive subordinate clause telling us only which class of numbers is being discussed. The phrase refers back to the introduction of the Ideas in Book V, where the Ideas are said to be each one of a kind rather than, like the phenomena, many of a given kind (476a).[8] The clause is simply telling us, then, that we are talking of Ideal numbers rather than of phenomenal numbers, not that the numerical identity of each Idea is what is under discussion.

This passage in the *Republic* (526a) asserts that the Ideas of numbers in respect to their internal content are each completely homogeneous in relation to each other; they differ not in the least from each other. For I take it that in saying "each unit" is "equal to all others," Plato is not vacuously asserting that "each thing that is numerically one is numerically one." "Each unit" here, in light of the subordinate clause, means "each Ideal number." And so Plato is making the surprising and seemingly paradoxical claim that some numbers, though they each differ in kind, nevertheless, in respect to their internal content, do not vary from one to the next. This suggestion of paradox would account for Plato here addressing the proponents *of his own view* as "strange people."

In addition, the passage claims that each Idea of number is internally homogeneous with respect to itself; each Idea of number has no parts. The Ideal numbers are not made up of any units, any merely numerical parts, and are not parts of each other.

The numbers discussed in the passage can hardly be mathematical intermediaries. For even if such intermediaries occupy the third level of the Divided Line (the level for objects of *diánoia*, thought, 510a–511d), the clause "in which the unit is as you assert" would still have no backward reference, since in the Divided Line passage no reference is made to the unitary nature or make-up of each occupant of that level (or indeed of any of the levels). Further, if Plato had wanted to say at 526a that he was intending to speak of numbers of an intermediary sort, made up of units which themselves, however, have no parts, he could have easily described his subject as "the kind of number (singular) . . . in which units (plural) are present." But he did not choose such a phrasing.

Plato does not tell us here (526a) what does differentiate the various Ideal numbers, given that internally they differ not even a bit from each

---

7 Cherniss (1) 515, 518, 521. For a criticism of Cherniss' position, see Allen (2) 91–93.

8 So correctly, Shorey 85.

other. Nor does he need to in order to accomplish his task of differentiating the two kinds of numbers: those which are the objects of reason and have no merely numerical units; and those which by contrast are the objects of opinion and guesswork (524e–525a) and do consist of units and have countable parts (525e).

Plato certainly does not suggest that the unitless numbers are the objects or products of arithmetical calculations (addition, subtraction, etc.). The Ideas of numbers are rather, like the other Ideas, objects of contemplative reason (525c2). On the other hand, Plato does not explicitly deny that the unitless numbers are the subjects of arithmetical calculations.[9] But unless we are to assign fortune telling powers to Plato that would have allowed him to see the claims made on his behalf about the numbers of *Republic* 526a and *Philebus* 56d–e, we should not expect him even to have entertained the problem, let alone to have answered it.

The *Philebus* passage, 56d–e, too, I will argue, refers to Ideas of numbers. To my knowledge only Shorey has seen this. Later critics who have braved interpretations of the passage uniformly maintain that it refers instead to mathematical intermediaries or mathematical numbers, that is, to objects of arithmetical calculation consisting of a plurality of units that are not, however, units of any particular kind. Annas calls such units "pure items"; Wedberg calls them "ideal mathematical units." Annas and Hackforth go so far as to assert that the *Philebus* passage constitutes the very best evidence there is that Plato did believe in mathematical intermediaries.[10]

We know, however, from the *Republic* what Plato means when he says that some units are equal units. The statement in *Philebus* 56e that "the unit belonging to each of the myriad [units that are numbers] differs in no way from any other unit [that belongs to a number]" is the counterpart of the statement in *Republic* 526a4 that equal units are numbers that have no parts or at least no merely countable or numerical parts. This interpretive move is not a mere argument from silence, but the spelling out of an abbreviation, given that we already know the full spelling. The alternative (taken by most critics) is to guess at what Plato means by the

---

[9] Cherniss seems to think that *Phaedo* 101b–c entails the denial that "the idea of two itself is not a combination of units or a product of factors." Cherniss (1) 517. Cherniss apparently deduces this denial from the statement in the passage that addition and division are not the causes of phenomenal twos. But I do not see how such an assertion about the phenomena can be claimed to hold *a fortiori* of the Ideas. One may as well argue that since phenomenal numbers have parts, so do Ideal numbers.

[10] Shorey 83n630; Wedberg 125–27; Annas (1) 6–7, 15, 20; Guthrie 230; Hackforth (4) 113n2, 115–16.

cryptic phrase "unequal units" taken in isolation, and then to interpret the "units that do not differ" in light of that (often uncharitable) guess. Even an argument from silence would be a preferable exegetical strategy.

Now I would argue that by "myriad" (μυρίων) Plato here refers to the fact that the kinds of different numbers are unlimited, that the number line is uncountably long. He is not referring to an unlimited source of constitutive units that make up numbers.

That the numbers of the philosopher's study have no merely numerical parts should be sufficient to indicate that these numbers are not mathematical numbers, which do have merely numerical parts. Again as in the *Republic*, Plato does not go on to tell us what does differentiate the various numbers that have no parts. But that these "equal" units are such in that they have no merely numerical parts gives us a sufficient clue for interpreting the problematic phrase "unequal units." To assign inequality to the numbers reckoned by the common man in contrast to the equality of all of the philosopher's numbers is simply, I suggest, to declare that phenomenal numbers do consist of countable numerical parts. The assignation of inequality here to phenomenal numbers does *not* mean either (1a) that the common reckoner uses distinct or different unit-concepts ("Before I was counting chairs, but now I'm counting parts of chairs") such that (1b) by contrast the philosopher always uses the same unit-concept (i.e., "ideal unit," "pure item") or (2a) that the counted units within the grouping or set are unequal in some respect other than their each being numerically one (say, by differing with respect to size), such that (2b) by contrast, the philosopher's units have no differentiation beside their each being numerically distinct (they have no qualities, for instance, or anything that would admit of degrees). Hackforth opts for alternative (2); Annas and Wedberg claim that Plato confuses (1) and (2) in our passage. I would argue that the confusion is foisted on Plato.[11] The inequality of the passage is not that of (1a) or (2a), and so Plato is not failing to distinguish them, and so too neither of the consequences put forward for the philosopher's numbers—(1b) or (2b)—follows.

If it is correct that the inequality here simply means that phenomenal numbers have countable parts, then the point of the enumeration of examples of unequal units (oxen, armies, etc.) all of the same number (two) is not that the common reckoner is compelled to use the same unit-concept for any one act of counting, but that even his countings and calculations are indifferent to unit-concepts, once any unit is established by reference to a unit-concept; he is capable of saying that two cabbages and two kings are each simply two, that one king and one cabbage and one

---

[11] Hackforth (4) 115–16, Annas (1) 8, Wedberg 125–26.

glob of sealing wax are collectively three things. The nature of twoness as immanent in the world is independent of being tied to any unit-concept, any particular kind of things counted.

If this is the case, then Plato is distinguishing numbers in the phenomenal world from the groups which they number. Annas, construing the *Philebus* passage along lines (1) and (2) above, maintains just the opposite, namely, that the passage is evidence for the view that Plato is not carefully distinguishing numbers from numbered groups.[12] But even in the *Republic* passage Plato distinguishes phenomenal numbers from the sensible and tangible bodies that they have (525d7–8) and so is distinguishing number from numbered group (see also *Theaetetus* 195e, and especially 198c2). Numbers for Plato fall in with moral and aesthetic terms, relations, and metaphysical notions as all picking out Forms whose instances in the phenomenal world are technically non-sensible, though such instances all supervene upon particulars that are sensible. Our "perception" of such instances requires judgments that use information provided only by an acquaintance with the Ideas or Forms of such instances (*Theaetetus* 184b–187c).[13]

Further, it is clear that Plato believes that numbers as immanent in the phenomenal world are perfect particulars of the Forms of which they are instances. By perfect particular I mean something that corresponds exactly to a standard. Such mathematical perfect particulars seem to be entertained at *Philebus* 51c and d.[14] In the *Phaedo* Plato makes it quite plain that phenomenal numbers are perfect particulars, for he says that numbers as immanent characteristics (ἐνόντων, 103b8), just like the Forms of which they are instances and from which they are derivatively named, do not admit of opposites (102a–107b, especially 104a–105b). The phenomenal number three cannot be partly another neighboring number, since it does not admit in any way of participation in evenness (see also *Cratylus* 432a–b). Now the story is different for the things that *possess* phenomenal numbers as characteristics. A folded letter may be both two and three in that it has two sides and three sections (*Republic* 524e–525a; *Philebus* 14d–e; cp. *Parmenides* 129c).

If this portrayal of phenomenal numbers is correct, it undercuts the largely *a priori* contention that, whether he said so or not, Plato must

---

[12] Annas (1) 8, cf. 4.

[13] See Cornford (1) ad loc.

[14] So correctly, but surprisingly, Hackforth (4) 99. Guthrie denies that the lines, circles, plane and solid figures mentioned there can be perfect particulars, for at 62b–c Plato allegedly calls such figures "false." Guthrie 226. But at 62b–c Plato is talking of figures arrived at by guesswork (στοχάσεώς, 62c1) rather than by reference to standards or measures, as I suggest is the case at 51c–d; and so Guthrie's objection is not apposite.

have believed in mathematical intermediaries, that he must have thought there exist some mathematical entities that are plural and perfect (or eternal as Aristotle claims, 991a4, 1028a18–21) such that the terms of arithmetical calculations like $2 + 2 = 4$ have referents.[15] The reasoning is that there allegedly can be no instantiations of arithmetical propositions in the phenomenal world, and yet the Ideas themselves cannot be objects of arithmetical calculation, since they are all *sui generis;* there is only one Idea of two. So, it is alleged there must be some *tertium quid* that is the object of arithmetical calculations. However, that $2 + 2 = 4$ is eternally true or true in every instance is the result of a fixed relation between Twoness, Fourness, Plusness and Equality. Further, that immanent numbers are perfect instances of their corresponding Ideas and neutral with regard to what the units are that make them up means that $2 + 2 = 4$ can have the phenomenal world as its referent, if it needs one; the proposition is elliptical for "two things and two things are four things," where "thing" means instance of *any* Form. And how many things there are taken severally (two here, two there) or collectively (four altogether) is determined by reference to the natural number line. There is then no need in the Platonic metaphysics to hypothesize the existence of mathematical numbers made up of "ideal units" or "pure items" in order to account for arithmetical calculations.[16] The role in arithmetic that most critics assign to the absolutely equal (philosophical) units of the *Philebus* actually should be assigned to the unequal (popular) units, which are neutral with regard to their non-numerical hosts. The equal numbers, which do not consist of units, are the Ideas of numbers.

If this account of the Platonic number theory is correct, it is handsome in the following way. If the Ideas of numbers have no parts, as I suggest is asserted in the *Republic* and *Philebus,* then they are not self-predicating. "Dozen" is predicated of an egg carton because the carton contains twelve units, but "dozen" is not predicated of the Idea of twelve in the same sense, if at all, for the Idea of twelve cannot consist of twelve units. Annas, overlooking *Republic* 526a altogether, supposes that Plato believed the Ideas of numbers to consist of units and so conceived of them as perfect instances of the characteristics that they are. In which case, the Ideas are subject to Aristotle's charges of being vacuous and vicious: vacuous because if the number two turns out to be just two units, then any two will do for Two (i.e., all twos will serve equally well as the number

[15] See Annas (1) 19, Hackforth (4) 115, and especially Wedberg 54–56, 68–71.

[16] It is important to note that *Phaedo* 101b–d does not (as Annas seems to suggest) deny that arithmetical calculations can have phenomenal objects as their referents; it is only denied that such processes are the *cause* of phenomenal objects having the numbers which they have. Annas (1) 20.

two), and vicious because subject to the Third Man Argument.[17] If my interpretation is correct, Plato avoids both of these charges. The Third Man argument is avoided, since there is no similar predication between an Idea of a number and its instances that would necessitate or even make possible the hypothesizing of a third thing to explain the (alleged) similarity of Form and instance. And the Ideas of numbers, as positions along the natural number line, are non-vacuous in two ways. First, the relations between such Ideas are able to account for the necessary entailments that hold between number concepts in a way abstractions from phenomenal particulars never can. And second, Plato would wish to claim that since instances of numbers are technically non-sensible, it is impossible to recognize even one instance of a number (let alone similarities between several instances) prior to possessing the content of the Idea of that number.

[17] Annas (1) 17. The Third Man Argument contends that if all Ideas have the very property which they cause in other things *and* if all properties are caused by an Idea over and above the property it causes, then for every type of phenomenal object there will be generated a vicious infinite regress of Ideas. See Plato's *Parmenides* 132a1–b2 and 132d1–133a6 and Aristotle's *On the Forms,* in Alexander (of Aphrodisias), *In Metaphysica commentaria,* ed. M. Hayduck (Commentaria in Aristotelem graeca, 1) (Berlin: Reimer, 1891), 84.21–85.11.

EXTENSIONS
(2005)

"Every human soul has seen, perhaps even before their birth, pure forms such as justice, temperance, beauty, and all the great moral qualities which we hold in honor. We are moved towards what is good by the faint memory of these forms, simple and calm and blessed, which we saw once in a pure clear light, being pure ourselves."

—Dame Judi Dench as Dame Iris Murdoch
creatively plagiarizing the
Phaedrus in *Iris* (2001)

SYLLABUS

I. Monadic *versus* Related Forms
II. Necessary Relations between Forms
III. Platonic Accounts
IV. Atomic Forms
V. Acquaintance with Forms
VI. Seeing Relations between Forms
VII. Knowing Forms as Individuals
VIII. Real Being
IX. Being and Goodness
X. Being as Being There
XI. God's Mixing Bowl: Some Final Thoughts on Divine Craftsmanship

In this coda, I will draw together, generalize, and extend some themes which appeared scattered through the book and which might gain from a consolidated and updated treatment. In particular I will address the extremely monadic nature of Forms as I've presented them in the first two chapters. I will chiefly be presenting extended metaphors that will provide analytical models for solving paradoxical elements of the interpretations that I've advanced. When I teach this material, I call the class sessions "the inner mysteries of Platonism." A clue: we will not be able to make sense of Platonism if our conceptual apparatus is set so that we can't help but think of Forms and their instances on an Aristotelian model in which for *any* thing that is there is something about it that makes it one in number and something else about it that makes it the type of thing it is, something that it shares as a common property with others of the same type. If Aristotle's distinction of matter and form is a mental lens through which we observe the furniture of the universe, then the truths of Platonism will elude our grasp.

I. MONADIC *VERSUS* RELATED FORMS

First, I should make clear that I believe that even in the middle, yes, even the early, dialogues, Plato thinks that Forms have significant relations of content between each other, that they are not all purely monadic with only merely formal, external relations (say, a relation of similarity with regard to oneness or eternality).[1] In the *Phaedo*, where the "no significant relations" view is usually thought to be grounded (in the "argument from the similarity of souls to Forms," 78b–84b), the Ideas of three and five are parts of the Idea of odd and the Ideas of two and four are parts of the Idea of even (103e–104b). And Forms are first introduced into the *Republic* with an assertion that there is a sense in which they are as determinate in each others' constitution as they are in the constitution of phenomena: "Each [of the Forms] is itself one, but because they appear everywhere in association with actions, and bodies, *and each other*, each appears to be many" (*Republic* V, 476a, Grube, my emphasis). Even in the *Euthyphro*, Socrates himself asserts that the Form of piety is a part of the Form of justice, and the Form of odd part of the Form of number (*Euthyphro* 11e, 12c–d). Significant relations of content between Forms are not an invention of the late, 'critical' dialogues.

On my account, though, each Form, in consequence of its being a standard, has some unique content that is irreducible to other Forms. Each Form has a core content that cannot be analyzed into other concepts. Each has a surd or root that is it itself alone. Each is fundamentally an individual with respect to what it is rather than a cluster of properties, or even just one single identifying property. But then the problem arises: if Forms can't be distinguished by properties, why aren't they fungible, substitutable for each other without difference. To save Forms from the Third Man Argument, haven't I imposed a restriction on them that makes them absurd?

Some intuitive sense, however, can be made of these curious conditions of Forms if it is noted that these conditions jointly apply to pre-theoretical, pre-Bohrian understandings of the referents of mass nouns (say, gold, glass, water, flesh, wood). Aristotle is on the mark when he gives these, rather than the referents of count nouns (horses, trees, golf balls) as paradigm cases of matter—that of which predications are made. They

[1] For the opposed view, see, for example, Nehamas (2) xliii.

are metaphysically distinct as that to which properties or forms attach, but are themselves neither simply properties nor form/matter composites. For each is distinctively a stuff. The referent of a count noun is not a stuff. A horse, a cow and a man are not three "stuffs" or three kinds of stuff. Aristotle is therefore off the mark with his doctrine of *relative* matters—the view that, for instance, gold is a qualification of yet another matter, water, in the same way that a ring's shape is a qualification of its gold. On this view, "gold" would become a count noun and gold would cease to be a stuff—would lose its most distinctive metaphysical character. A form/matter or attribute/thing analysis applies neatly only to the referents of count nouns.

Indeed common understanding does not suppose that the referents of mass nouns are *things* which have properties that account for their various distinctive natures. Rather the properties which common understanding associates most closely or "essentially" with the referents of mass nouns are properly ascribed rather to things *other* than the referents themselves—things which, however, have the properties by virtue of the referents themselves serving loosely as standards for identifying the properties. Thus one casually says that water is wet, perhaps even thinks that wet defines water, but really what one means by wet is that a wet thing is one that has water on it or in it in a way that can be felt. One actually uses water to identify things as wet, not the other way around. Or one says that water is drop-forming, but really what is meant when one calls something drop-forming is that it acts in air as water does. Or one says that water is life-giving, when what one really means is that, for a non-artificial object, taking in water is required to maintain it as the kind of thing it is. So similarly stand to each other the items of the following couples: gold/golden, lead/ leaden, flesh/fleshy.

On the other hand, everyday pre-theoretical, pre-Bohrian understanding does not suppose that some deep, hidden, ultimately quantifiable structure or "genotype"—say, H_2O or the 79 electrons, 79 protons, and smattering of neutrons that make up Bohrian gold—is metaphysically lurking in the nature of things to guarantee and account for the unique natures of the referents of mass nouns.

Common understanding tends to view the referents of mass nouns as primitives. Water is a prime case. Once it is seen that the properties most closely associated with it are not what make it what it is but are consequences in other things of what it is, nothing is left to say about it in the language of properties that gets to the heart of what it is. And yet it is different than gold. The unique content of the referent of a mass noun is neither analyzable properties nor fungible particularity. Something like this understanding of mass nouns as indicating primitive contents, I suggest, stands behind Plato's understanding of the unique contents of Forms.

II. NECESSARY RELATIONS BETWEEN FORMS

This understanding of mass nouns also stands behind Plato's understanding of the necessary relations between contents of Forms. The descriptions in the late dialogues of relations between contents of Forms tend to be elaborated in terms of metaphors of artistic and natural production which apply chiefly and in some cases exclusively to the referents of mass nouns: mixing, blending, intermingling; pervading or running through; purifying or cleansing; interweaving; harmonizing; cutting or chopping up.[2] These are metaphors of processes by which stuffs come to be related or distinguished and help explain how Plato can both maintain the view that the contents of Forms may necessarily entail each other and yet that no Form's content is exhaustively analyzable into that of other Forms.

The relations between contents of Forms may be viewed, as Plato's metaphors suggest, on an analogy with blendings which produce alloys. Brass would not be what it is if copper and zinc were not what they are, and yet one has not exhausted or even much clarified the nature of brass in saying that it is copper blended, pervaded, mixed, or infused with zinc. Such an account of brass fails, in our pre-Bohrian world, to explain, for instance, the look or presence of brass. An unanalyzed surd remains. Further zinc and copper are not said of brass, nor brass of either of them. Neither of them is either a genus or a differentia of brass, and yet they, in the distinctive ways of stuffs, are as "essential" to it as anything might be. So too, I suggest, Plato views the contents of Forms as "blending" and standing to each other in the ways that the referents of mass nouns are distinctively interrelated. They may stand in relations of necessary entailments to each other and yet not constitute singly or in groups the "essence" of each other. In relations between Forms, analyticity and necessity come apart. There will be necessarily true statements of Forms that are not analytic truths ("brass is zinc plus copper").

III. PLATONIC ACCOUNTS

It has been claimed that the extreme monadism of Forms that I advance will destroy knowledge for Plato because he incorporates into his understanding of what counts as knowledge—what distinguishes it

[2] To take just the *Sophist*: mixing, blending, intermingling, 251d5, 252b6, e2, 253b9, c2, 254d7, d10, e4, 259a4, 260b1; pervading or running through, 253a5, d6, 255e4, 259a6; interweaving, 259e6; harmonizing, 253b11; cutting or chopping up, 219d9, 227d1, 257c7, 258e1. For purification, see the *Statesman*'s extended metaphor of smelting (303d–e).

from merely having true opinion—the ability of the knower to give a λόγος, an account, of what the thing is that is known.[3] My reading of Forms might seem to rule out giving accounts of what they are. I wish to suggest, though, that Plato has a very loose notion of what counts as an account for these purposes.

First, the meaning of λόγος itself is quite elastic. Here are the entries for it from *LSJ*: "Thought either as expressed or as residing in the mind. *I.* what is said: sentence, proposition; saying, maxim, proverb; assertion, promise; a command; speech, discourse; talk which one occasions (repute); a tale, story; prose, prose-writing; principle; definition. *II. ratio,* power of mind: reason; an opinion; ground, plea; account, consideration, esteem; count, reckoning; tale; reckoning, account, bill; due-relation, proportion." It would almost appear that anything short of a person being completely tongue-tied will do for giving a λόγος. Surely something far short of what we would count as a formal definition will fit the bill of being a λόγος. And in the *Timaeus* when Plato actually gives us an account about Forms that is as certain as accounts can be, the account in fact is not a formal definition, but something much grander; it is a statement of a six-term relation between Oneness, Twoness, In-ness, Difference, Simultaneity, and Becoming: "On the other hand, that which has real being has the support of the exactly true account, which declares that, so long as two things are different, neither can ever come to be in the other in such a way that the two should become at once one and the same thing and two" (52c7–d1, Cornford, slightly altered). Here to give a λόγος of a Form is to trace its relations to other Forms. So giving a λόγος for the purposes of giving evidence of having knowledge can be both much less and much more than giving a defintion.

Second, in his actual practice of giving definitions more narrowly understood, Plato does not require that an account (λόγος) that refers to the content of a Form be given in the Aristotelian form of a genus plus differentia—a form in which the concept is exhaustively analyzed into other concepts. Indeed in the *Theaetetus* Socrates gives as a model definition, a definition that appeals both to the thing defined and the parts of the definition as mass nouns. He defines clay as earth mixed with water (*Theaetetus* 147c). Neither earth nor water is either a genus or differentia of the other. As with my brass example, Socrates' definition of clay is as informative about clay as anything is, but it does not get to the heart of what clay is. It does though distinguish it from all other concepts, and, importantly, does so without appeals to what clay does or suffers. Platonic definitions trace the mass-noun-like relations that exist between

[3] The classic formulation is given at *Meno* 98a3–8; see also *Symposium* 202a5–7, *Philebus* 62a3.

forms without appeals to processes. To appeal to doings or sufferings is to state a quality or condition of a thing not its 'essence' (*Euthyphro* 11a, *Meno* 71a–b, *Republic* I, 354b–c, *Phaedrus* 270d).

So we actually end up three levels of Platonic "essences": 1) each form will have a core content that is non-analyzable; 2) most forms will have substantive relations with other forms which are like relations between the referents of mass nouns—one can give an account of a Form by tracing these relations in propositions; 3) for some Forms necessary and sufficient conditions may be given which appeal to the Form's doing or undergoing something; these accounts too can distinguish one Form from all others, but are not to be settled for if discursive accounts of the previous type are possible. So in the *Meno*, Socrates prefers the type of definition exampled by the definition of shape as "a limit of a solid" (76a) to that exampled by the definition of color as "an effluvium from shapes which fits the sight and is perceived" (76d–e).

As examples of all three levels of Platonic "essences," consider brass again. It will have a core essence that is its look, one distinct from bronze. It will also have a discursive essence—copper plus zinc—which also distinguishes it from bronze (copper plus tin), but which does not fully explain its look. And both brass and bronze will have qualities or affections that are distinctive of each, but which do not properly define either. Bronze is the metal that works best for making sculptures and that is most loved by the Greeks, while brass is the metal that is best for making bells and tumbler locks—and is most loved by the Persians.

Finally, even if one cannot give a discursive definition of a Form, Plato has a still looser understanding of what counts as giving an account of a Form for the purposes of giving evidence of having knowledge. In this understanding, to give an account of the Form is simply to be able to retrace discursively for others the steps one took in getting to the point of "seeing" it. This giving of a λόγος tells us nothing about the internal content of a Form. The course that is retraced can be a route from Form to Form, or from phenomena to Form, or even just be one's general thought patterns. Since this sense of giving an account in Plato has all but been overlooked in the literature, let me give several examples.

In the passage of the *Phaedo* that bridges the "argument from recollection" and the "argument from the similarity of souls to Forms," Socrates gives the canonical claim that the person with knowledge is the person who can give an account:

> A person who has knowledge would be able to give an account of what he knows, or would he not?
>
> He must certainly be able to do so, Socrates, he said.
>
> And do you think everybody can give an account of the things [i.e., Forms] we were mentioning just now?

> I wish they could, said Simmias, but I'm afraid it is much more likely that by this time tomorrow there will be no one left who can do so adequately. (*Phaedo* 76b)

But only two pages later Socrates says the following about giving an account:

> Let us return to those same things with which we were dealing earlier [i.e., Forms], to that reality of whose being we are giving an account (αὐτὴ ἡ οὐσία ἧς λόγον δίδομεν τοῦ εἶναι) *in our questions and answers. . . .* (*Phaedo* 78c10–d2, Grube adapted, my emphasis)

It turns out that simply in their back-and-forth discussion, Socrates and Simmias have fulfilled the discursive requirement for knowing, the ability to give a λόγος of a thing, and yet they have not given any definitions of any Forms in their questioning and answering. Indeed they have not given for *anything* a definition whether by genus and differentia or any other exhaustive and distinctive analytical account of a concept. Simply recounting their path in discussing Forms is sufficient to meet the requirement of giving an account.

In the discussion of the Form of the good in the *Republic*, it appears that simply retracing the dialecticians ascent up the Divided Line counts as giving an account for the purposes of establishing that one has knowledge of the Good—retracing the shift from focusing on many good things to the Good itself is sufficient for the discursive requirement for knowledge (534b):

> And you also call a dialectician the man who can give a reasoned account (λόγος) of the reality (οὐσία) of each thing? To the man who can give no such account, either to himself or another, you will to that extent deny knowledge of his subject—How could I say he had it?
>
> And the same applies to the Good. The man who cannot by reason (τῷ λόγῳ) distinguish the Form of the good from all the others, who does not, as in a battle, survive all refutations, eager to argue according to reality (οὐσία) and not according to opinion, and who does not come through all the tests without faltering in reasoned discourse (τῷ λόγῳ)—such a man you will say does not know the Good itself, nor any other good. (*Republic* VII, 534b3–c5, Grube)

Here we know the Good not by distinguishing it definitionally from other concepts but by being able to distinguish it from other good things, the many goods. Reasoned discourse (τῷ λόγῳ) is a *path to* the distinction between the Good itself and the many good things and to be able to retrace this path is sufficient for the purpose of giving an account as part of what it mean to have knowledge of the Good. Nothing here suggests that the knower will be able to give a defintion of Good by genus and differentia or in any other discursive form that might count as a definition.[4] In-

---

[4] Reeve has a different and incompatible reading of this passage on this point. Reeve 72, 85, 91–92.

deed Plato immediately re-deploys his dreaming-waking analogy to describe the process coming to have knowledge of the Good. To know the Good is like waking up from a dream whose objects are merely many good things and which are mere images of the Good itself (534c5–d1). One could say something about this transition—Plato just has—but we should not expect a description of the Good at the end of the process.

In the *Timaeus* Plato telescopes these claims from the *Republic* into the claim that Forms are apprehended "by thought with the aid of an account" (νοήσει μετὰ λόγου, 28a1–2). Λόγος again is what puts one on the way to Forms. When Plato later says that one of the things that distinguishes knowledge from opinion is that knowledge "is always with a true account" (τὸ μὲν ἀεὶ μετ' ἀληθοῦς λόγου, 51e3), he is surely referring back to 28a1. The requirement then is not to give a definition or any discursive account of the content of a Form, but simply to be aware of and to be able to articulate the process of getting to it.

In the *Phaedrus*, in the passage that links the philosophical methods of collection and division with the doctrine of recollection (265d), it is said that we have the capacity to define a term simply through the process of collection—division and analysis are not necessarily involved—and half of the things that count as collections it will turn out are recollections of Forms:

> The first philosophical method consists in seeing together things that are scattered about everywhere and collecting them into one kind (ἰδέα) so that by defining (ὁριζόμενος) each thing we can make clear the subject of any instruction we wish to give. (*Phaedrus* 265d3–5, Nehamas)

The defintion-generating collection entertained here can have the form of a shift from specific Forms to a generic Form, but thanks to the process of recollecting past contacts with Forms, it can also take the form of shifting from phenomenal instances of a Form to that Form itself, as Plato makes clear in the passage to which *Phaedrus* 265d is a backwards reference:

> But a soul that never saw the truth [that is, at least one Form] cannot take a human shape, since a human being must understand speech in terms of Form, proceeding to collect together by reasoning many perceptions into a unity. That process is the recollection of the things our soul saw when it was traveling with god, when it disregarded the things we now call real and lifted up its head to what is truly real instead. (*Phaedrus* 249b5–c4, Nehamas adapted)

The example Socrates has in mind here is the soul collecting together many perceptions of beautiful objects and having that process trigger a recollection of the Form of beauty as an isolated unity. The process of isolating the Form of beauty by recollection, we are told by the latter passage (265d), counts formally as having defined what beauty is. Neither passage entails that the knower with his term defined is going to be able to give a discursive account of its content.

And this is good news for Platonism. Because there are two passages in Plato that seem to assert the existence of Forms which are ultimately simple, which have no parts and are not parts of others. That a description of the process of coming to (know) a Form counts as a kind of "giving an account" that is sufficient evidence of possessing knowledge saves these Forms from being unintelligible.

IV. ATOMIC FORMS

The two passages in Plato that directly suggest that some Forms are completely unanalyzable with respect to their content are somewhat technical. At *Phaedrus* 270d1, some Forms are simple (ἁπλοῦν) as opposed to complex (πολυειδές). Such Forms do not have parts that can be "enumerated," placed in a rational ordering.[5] These are Forms which do not have species. If they are said of other Forms, they are said of each in the same way, unmediated by differentiae. When Plato goes on to say that nothing can be said of them other than to describe their capacities or powers (δύναμις, d4), it is also implied that they are not articulately differentiated parts of other Forms. They neither have species nor are species of other Forms. The capacities here are the natural ability to act upon a thing (πρὸς τί πέφυκεν εἰς τὸ δρᾶν) and the natural ability to be acted upon by something (εἰς τὸ παθεῖν ὑπὸ τοῦ, 270d4–5). An example of the latter, a Form being acted upon by something, is piety being loved by the gods (πάθος, *Euthyphro* 11a8), and examples of a Form acting upon something are the Forms of fire and water igniting and liquefying parts of the receptacle of space by casting their images upon it (*Timaeus* 51b4, 5, 52d5).

---

[5] This passage fine-tunes the discussion of collection and division at 265d–266b, which in turn refers back to the discussion of collection, Forms, and recollection at 249c—as even those who wish to deny that the theory of Forms operates in the second half of the *Phaedrus* (257c ff.) self-contradictorily admit. Nehamas (2) 64n149.

The Forms in question in the passage are types of soul or what we would call personality types or characters. For Plato in the *Phaedrus* personality types are treated as natural kinds, not arbitrary social conventions (248a, 248d–e, 252c–253b), and so are as subject to analysis in terms of Forms as are the good and the beautiful. Collection and division have not become empirical or sociological methods. Their empirical status is explicitly denied one page earlier—at 270b: "In both cases we need to determine the nature of something—of the body in medicine, of the soul in rhetoric. Otherwise, all we'll have will be an empirical and artless practice" (Nehamas).

Call those Forms that neither are parts of other Forms nor have other Forms as parts "outliers." Such outliers are asserted as one of the four divisions of Forms in the *Sophist*'s grand summation and schematization of relations between Forms: "Finally, [the dialectician understands that there are] many Forms that are separate in every way from others" (253d9). These are Forms whose contents are not linked to the content of any other, but which do share formal or external properties (such as, being, unity, sameness, rest) with other Forms.[6] In neither the *Phaedrus* nor the *Sophist* does Plato actually give an example of a Form that falls in this category, but these Forms wholly exemplify something that is true of each Form. Each Form will have an aspect of its content that falls below the radar of definition, even as most of them will have definitions, broadly construed, that is, there will complexes of other Forms properly said of them in virtue of their content, complexes that tell us informative things about them but that are not merely a recounting of the Forms' "capacities."

V. ACQUAINTANCE WITH FORMS

The press in Platonic scholarship to turn Platonic knowledge into something purely discursive has become ever more intense over the last two decades, leaving almost quaint my 'Neoplatonic lite' claim that for Plato knowledge is a form of direct acquaintance, a seeing of the Forms with the mind's eye, and worse than that, that this direct acquaintance is a form of seeing by which we take in individuals as individuals, not as we take in repeatable properties, like colors.

It is worth remembering, though, just how extensive the metaphor of knowing as seeing is in Plato. Here is an incomplete list of cases where Plato uses "seeing," and direct acquaintance more generally (including moving toward an object), as his master metaphor for knowing—from the *Euthyphro* straight through to the *Philebus*:

Euthyphro 6e4 (ἀποβλέπειν)

Greater Hippias 299e2 (ἀποβλέπειν)

Meno 72c8 (ἀποβλέπειν)

Cratylus 389b2 (βλέπειν)

Symposium 210d4, e3, e4, 211b6, b7, d2, d3, e1, e4, 212a2, a3, a3.

Phaedo 65e7, 66a3, d7, e1, 74c9, 79a3, d6, 84b1, 109e6.

[6] These formal attributes of Forms are each said to be "a single Form extending all through the many Forms each of which lays separate from the others" (*Sophist* 253d).

Republic VI, 490b3, b4, b5, b5, 500c3, c3, d4, 501b1 (ἀποβλέπειν), 508d4 (eye of soul); VII, 517c1, c1, c5, 518c6, d6, 519b3, b5, 533d2 (eye of soul), a7 (eye of soul), 540a8 (ἀποβλέπειν).[7]

Phaedrus 247c7, d3, d4, d5, d6, d6, e3, 248a4, a6, b4, b6, c3, c6, d2, 249b6, c2 d7, 250a2, a4, b5, b6, b8, c4, 251a2, 254b6, b7.

Timaeus 28a7 (βλέπειν), 29a3 (βλέπειν).

Sophist 254a10 (eye of soul), b1.

Philebus 61e1 (ἀποβλέπειν).

This catalogue is a lot for *Plato discursus* critics to explain away. In addition, the direct-acquaintance metaphors are both quite specific and elaborately extended, conditions which make Platonic seeing unlike the passel of dead metaphors of direct acquaintance used in contemporary English to indicate knowing, for example, "I see what you mean," "I grasp your meaning," "I get your point," "I hear you, buddy," "I feel your pain."[8]

Specific. Among the host of terms for seeing that Plato uses as metaphors for the mind's taking in of Forms, his term of choice is ἀποβλέπειν, which means to look concertedly at one object, "to look away from others to one thing; to gaze steadfastly; to look in a particular direction; to look upon with love, wonder, or admiration; to look longingly" (*LSJ*, s.v.). Thus Socrates requires Euthyphro to tell him "what this Form (ἰδέα) itself is, so that I may look upon (ἀποβλέπειν) it and using this model (παράδειγμα), say that any action of yours or another's that is of that kind (τοιοῦτον) is pious, and if it is not that it is not" (*Euthyphro* 6e3–6; and see *Greater Hippias* 299e2, *Meno* 72c8). This is the verb used of the mind's contact with the Forms of virtues viewed as divine paradigms in *Republic* VI (501b1) and with the Form of the good taken as a paradigm in *Republic* VII (540a8, cf. 484c9, 532a4). In the *Philebus,* it is the verb used to express how the mind's contact with what is always and unchanging constitutes full knowledge (61e1, cf. 44e2, 45a1). In its root form, βλέπειν, the term is used in the *Cratylus* for a carpenter's looking to the Form of shuttle as the model for his makings (389b2) and in the *Timaeus* for the way the craftsman God looks to the Forms to use them as models for his makings (28a7, 29a3).

Extended. The metaphors of direct acquaintance in Plato are complex—which is to say they offer explanatory models. Used, as they are, in conjunction with Forms taken as standards and originals and with the additional metaphor of "the eye of the soul" (*Republic* 508d2, 533d2, *Sophist* 254a10), these occurrences of "to gaze concertedly with admiration" are

---

[7] Not to mention all the dreaming/waking contrasts in *Republic* V–VII, 476c, 533b–c, 534c, and blind/sighted distinctions, 484c, 506d.

[8] Contra Gosling (1).

extended metaphors and cannot be explained away as being on a footing with the way we say, "Oh, I see what you mean," where nothing visual or visual-like is conveyed by the metaphor of sight.

We can have direct acquaintances with simples even if we cannot describe them. Alexander Nehamas has noticed that Plato's metaphors of sight are especially apt if Plato is seeking a mode of cognition for simples: "Plato often describes our coming to know the Forms, as he does in the *Phaedrus*, by metaphors drawn from sight. Such metaphors reinforce the notion that each Form is an independent object which can be seen in its entirety by itself."[9] But Nehamas then ought to be distraught that Plato continues to use metaphors of vision for models of knowing right on into the late critical dialogues, where Nehamas supposes that Plato has given up both the monadic nature and transcendental status of Forms.[10] The dialectician of the *Sophist* knows of the simples that are outliers; it's one quarter of his job to recognize them. And though it is hardly ever remarked, a complex extended metaphor of vision is also advanced in the *Sophist* as a model of knowing. Immediately after the summation of the dialectician's taxonomy of Forms, we are told that "The philosopher always uses reasoning to stay near the Form of being. He isn't at all easy to see because that area is so bright and the eyes of most people's souls can't steadfastly have the divine in full view (πρὸς τὸ θεῖον ἀφορῶντα)" (254a8–b2). In its use of the paradoxes of light, its mention of the eye of the soul, its assigning divinity to the Forms, and its central metaphor of knowing as a kind of concerted gazing, the passage is clearly meant to echo the Cave and Sun analogies in the central books of the *Republic*.

VI. SEEING RELATIONS BETWEEN FORMS

While forms have necessary relations between each others' contents, these relations are synthetic not analytic. One does not automatically "get" a form that is necessarily related to a Form that one already knows. Rather the Forms hint at or suggest or point to each other. Relations between Forms lead the mind's eye from one Form to the next in the way that the parts of a well-composed painting lead the eye around the canvass. Just to mention two famous examples. In Vincent van Gogh's *Irises* (1889) there is a general sweep of visual interest running from the lower right corner toward the upper left. The eye runs almost as though in grooves along the flower stalks that have been partially knocked down by the wind. The sweep of the eye is so strong that it would run right off the canvass onto the wall except that right at the last moment of the sweep there is one startling large white iris among the fully colored ones.

[9] Nehamas (2) xlii.

[10] Nehamas (2) xliii–xlvii.

It catches and arrests the eye, but then pressing at the right edge of the painting, there is a smaller, partially white bloom that then catches the eye from the large one and draws the eye across the tangle of blue irises back to a point where the eye can again be caught up in the general diagonal sweep across the canvass. The various parts of the picture suggest themselves to each other. Even more well known are the overlapping obtuse triangles that form large scale structures within the chaos that is Picasso's *Guernica* (1937). The triangular forms with their long bases paralleling the painting's bottom edge at once unify and intensify the painting's jagged mid-sized forms and lead the eye around to every part of what is a very busy painting.

In the *Phaedrus*, Plato takes us to another sort of museum to provide a metaphor for the process of various Forms suggesting each other to the observer's mind. We are in the interior of a temple where Forms have been placed upon sacred pedestals where normally cult statues of the gods would stand (254b7). The context: When the lover sees the beauty of his boy, that beauty has a double, almost contradictory, effect on him (254b–c). He is inflamed with passion and wants, even sets in motion, to ravage the boy sexually, but the boy's beauty also causes him to recollect the Form of beauty, which he had seen in an earlier bodiless life, and contact with that Form in turn puts him in contact with the Form of self-control: "So they [the lovers] are now close to him [each their boy], and they are struck by the boy's face as if by a bolt of lightning. When the charioteer sees that face, his memory is carried back to the real nature of Beauty, and sees it again where it stands on its sacred pedestal next to Self-Control. At the sight, he is frightened, falls over backwards awe-struck, and at the same time has to pull the reins back so fiercely that both horses are set on their haunches" (254b–c, Nehamas). The Form of beauty suggests to the mind that perceives it the Form of self-control or temperance, and the latter, when perceived, has the effect on the lover's soul of reining in his lust for the boy's body. Beauty is not analytically entailed by temperance or vice versa. For example, beauty never even comes up in the definition or discussion of temperance in Plato's most extended discussion of temperance and the other virtues, in *Republic* IV (430e–432a, 442c–d). But the two evaluative Forms—Beauty and Moderation—suggest each other and lead the mind's eye from one to the other. This is the path of dialectics, beginning with Forms moving by means of Forms and through Forms to still other Forms, and finally ending in Forms (*Republic* 511c).

VII. KNOWING FORMS AS INDIVIDUALS

In 1985, I was operating on a hunch when I wrote that we take in individual people, though not other things, directly as individuals without analytically distinguishing features or properties of them which when taken collectively would mark them out as each unique. This hunch now has the imprimatur of literary history and biology.

Literary history. In chapter six of Lewis Carroll's *Through the Looking Glass* (1871), the conversation between Alice and Humpty Dumpty grinds to an awkward halt. Alice tries to extricate herself politely from the scene, but is unable to do so without Humpty Dumpty crossing metaphysical swords with her:

> "Is that all?" Alice timidly asked.
>
> "That's all," said Humpty Dumpty. "Good-bye."
>
> This was rather sudden, Alice thought: but, after a *very* strong hint that she ought to be going, she felt that it would hardly be civil to stay. So she got up, and held out her hand. "Good-bye, till we meet again!" she said as cheerfully as she could.
>
> "I shouldn't know you again if we *did* meet," Humpty Dumpty replied in a discontented tone, giving her one of his fingers to shake: "you're so exactly like other people."
>
> "The face is what one goes by, generally," Alice remarked in a thoughtful tone.
>
> "That's just what I complain of," said Humpty Dumpty. "Your face is the same as everybody has—the two eyes, so ______" (marking their places in the air with his thumb) "nose in the middle, mouth under. It's always the same. Now if you had the two eyes on the same side of the nose, for instance—or the mouth at the top—that would be *some* help."
>
> "It wouldn't look nice," Alice objected. But Humpty Dumpty only shut his eyes, and said "Wait till you've tried."

Of course, our heroine wins the metaphysical contretemps because indeed it is the face that one usually goes by when individuating people, and we do it, as Alice suggests, without the sort of analysis of features that Humpty Dumpty insists upon, and, of course, too Humpty Dumpty himself cannot be constructed as the individual he is from his divided parts ("All the King's horses, And all the King's men / Couldn't put Humpty together again!"), a fact to which Carroll alludes only obliquely at chapter's end: ". . . a heavy crash shook the forest from end to end." Out of Alice's visual range, Humpty has had a great fall.

One of the two terms that Plato uses for "Form"—εἶδος, deriving from *εἴδω—has as its root sense "that which is seen." Eventually the term becomes quite abstract, meaning "sort, particular kind or nature, class, kind, species." But in its earliest uses it retains and even particularizes its root sense by referring specifically a *human* form or figure, the *look*, the good looks, of a person (*LSJ*, s.v.), their 'visage', as it were. And this use continues on into Plato. In the *Phaedo*, Plato gives, as his first example of the processes of recollection, the case where a lover sees the lyre or a garment of his boy and is thereby reminded of the boy's εἶδος, his particular look (73d8, cf. *Protagoras* 352a2, and espcially *Charmides* 154d4). I suggest that this sense of εἶδος, as the looks of a particular individual, is not lost in Plato's metaphysical uses of the term, but remains a strong pentimento presence in the senses of the term when it means "Form."

Biology. It turns out that the brain recognizes and identifies faces through a different pathway than it identifies properties—color, depth, movement, even facial expressions. This division in cognition was discovered by the study of people who have a rare brain disorder called prosopagnosia, a syndrome in which people lose the abilty to identify faces, including faces of parents, children, spouses, even their own faces in mirrors. The disorder is caused when lesions appear in two small matching areas of tissue deep in both sides of the brain. The lesions can be caused by stroke, head injury, or a form of encephelitis. It is the syndrome made famous by Oliver Sacks' book *The Man Who Mistook his Wife for a Hat*. People with the disease can be as adept as people with normal vision in assessing expressions, age, and gender, but cannot identify faces.[11] This distinction in humans' cognition of humans provides an analogy for how Plato can use sight as his master metaphor for how we can take in Forms as individuals, how we can 'perceive' Forms without Forms being understood as composites of properties, or even as each a single simple property with respect to what it is.

There is knowing *that* . . . such and such is the case. There is knowing *how* . . . to bake a cake. And there is knowing *who* . . . , Mom. For Plato, just as "knowing *that*," when said of the phenomenal realm, is thoroughly conditioned by "knowing *how*" to get along there, so too "knowing *that*," when said of a Form, is thoroughly conditioned by an analogue to "knowing *who*" in particular someone is.

VIII. REAL BEING

One critical trend that has continued to gather force over the last twenty years is the ever wider acceptance of Charles Kahn's view that the Platonic "to be" is not an existential sense of "to be"—"to exist"—but rather a predicative sense—"to be F" or "to be some complex of F, G, H, . . . ," even when the verb is used without any complement.[12] Kahn claims that the sense of "to be" meaning "to exist" does not come fully into its own in ancient Greek, does not really become a free-standing usage. Now if one believes that all knowledge is propositional and if one is

---

[11] Sandra Blakeslee, "Study of Rare Syndrome Yields Clues about Brain," *New York Times*, May 24, 1988, p. 23.

[12] Kahn (1), (2) . Lesley Brown has rearticulated and developed Kahn's view. For new adherents, see, e.g., Burnyeat 225n31.

trying to make Plato respectable, it is understandable that one would follow Kahn's lead in having Plato committed to a predication-saturated universe. And the move to expunge existence from the Platonic palette has the added bonus of having Plato not being able to speak of modes or degrees of existence. Dualism is made impossible, all the more so any Neoplatonic proliferation of grades of existence. I can't here go thoroughly into the issue of whether Kahn is right about Plato since to do so would involve too many complex texts, especially the much disputed uses of "to be" that accompany the introduction of Forms at the end of *Republic* V (476a–480a). But I do want to suggest that it is a view that does not fit the *Timaeus* very well.

Take the very opening of Timaeus' discourse. Timaeus' first question is whether the cosmos has come-into-being or whether it is ungenerated (27c5). Half a page later, this question is rephrased as the question, "Has it always been?" (ἦν ἀεί, 28b6). The "be" verb here means "to exist" and nothing more. What is being probed is "Whether the fabricated universe always existed?" Even Zeyl admits this: "The question whether the world 'has always been, having no origin of its coming to be or [whether it] has come to be, taking its start from some origin' (28b6–7) does seem to be the question of whether the world is something that has always existed . . . . We should be careful, then, not to exclude existence from the semantic range of this verb in this opening section of the discourse."[13]

But Zeyl, then, oddly turns right around, by referencing Kahn, to exclude "exist" as the sense of "to be" used in the explanation of the grand distinction between being and becoming—the distinction between the mode of Forms and the mode of the phenomena—which is drawn *between* the two formulations of the question about whether the cosmos has always existed (27d6–28a1).[14]

After making a meta-level nod to the gods (27c–d), Timaeus begins his discourse proper: "As I see it, then, we must begin by making the following distinction: What is *that which always is* and has no becoming, and what is *that which becomes* but never is" (27d6–28a1, Zeyl). If the "be" verbs here are understood predicatively, then a Form will be that which is always F, and never comes to have nor ever loses the property F, or in Zeyl's phrasing: "To say, then, that something 'always is' is not as such to say that it exists for all time but just as readily to say that it always is just what it is, something F or G," while an instance of the Form is something which becomes F, but never is F—or to be coherent, never becomes fully F, or in Zeyl's phrasing: "By the same token to say that something 'never is' is not as such to say that it never exists, but that it never is F, for any value of F—that it fails to possess any definite characteristic . . .

---

[13] Zeyl (2) xxx–xxxi.

[14] Zeyl (2) xxx.

moving, perhaps *toward* being F or G but never completing the movement."[15] This position requires that there can be no perfect particulars of Forms in the phenomenal world, otherwise the particular would fully have the property which the Form alone is supposed to have fully and the distinction between Form and particular collapses as far as their contents go, though the Form is longer lasting than the phenomenal object. Unfortunately for Zeyl's reading, there are perfect particulars in the Timaean world. At a minimum, at least some of the instances of the Forms of the four primary bodies (earth, air, fire, water) are perfect particulars, just exactly are instances of four of the five Platonic solids, the perfect details of which are minutely worked out by Timaeus at 53c–55c. Further since there are, even in the pre-cosmos, multiple objects in the receptacle of space, groupings of these will perfectly instantiate Forms of integers, which both in themselves and in their instantiations do not admit of degrees (*Phaedo* 103e–105c, *Cratylus* 432a). The Form of motion too will have perfect instances in the pre-cosmos; everything there is moving around to the greatest degree possible. And in the demiurgically sired world the World-Soul looks to be a perfect particular in its perfectly unerring revolutions (47c4).

The "be" verbs of Timaeus' opening distinction between being and becoming then are not predicative uses, but existential ones, as we should expect from Timaeus' question about the *existence* of the demiurgically generated world. Plato's existential use of being here marks out different orders or levels of existence. Traditional, heavily dualistic interpretations of the opening of Timaeus' discourse are correct. The second half of the *Timaeus* goes on to explain that indeed the flux of the phenomena draws the very existence of the phenomena into doubt because flux tends to sap the phenomena of any presence in consequence of sapping them of any essence or whatness. But the phenomena are saved from utter non-existence by their status as images appearing in the medium of space. They have a lower grade of existence than the Forms do. For they are dependent upon other things for their ability to shine forth. Forms do that on their own, but the phenomena can do so only as the result of relations they hold to both the Forms and the medium of space. The Forms therefore fully exist, while the phenomena are less than fully real. They hover between what fully exists and the void, caught from the plunge into (becoming) nothingness by the medium of space.

IX. BEING AND GOODNESS

Kahn's existence-barring, predicate-requiring reading of "to be" also fails to fit Plato's discussion of the Form of the good and its relation to

[15] Zeyl (2) xxx, Zeyl's emphasis.

being and becoming in *Republic* VI (509b). The Form of the good is said to cause the other Forms to be. The Good provides them τὸ εἶναί τε καὶ τὴν οὐσίαν (509b7–8). Does this mean, as Kahn's reading would require, that the Form of the good causes the Form of two or Form of courage to have the properties that make each the particular Form it is? The metaphor of the sun seems to rule this out. Plato offers the analogy that the Form of the good stands to the being of the Forms as the sun stands to the becoming of things that become. He seems to have vegetal becoming particularly in mind: the sun provides the objects of sight with "their generation, increase, and nurture" (509b3–4). Now, we might ask: does the sun, its light, which is itself undifferentiated, cause some undifferentiated stuff to become soy bean fields, even as it takes some other portion of undifferentiated stuff and makes it into corn fields—or does sunlight cause things that are already determinately soybeans and corn, say, soybean sprouts and corn sprouts, to grow. The answer is obviously the latter. The sun helps realize the becoming of things that become, it realizes their status, but it does not cause differences in kind between the things with that status; those differences have a different source, say, soybean seeds and corn kernels. But then when we switch back to consider Forms, the sun analogy means that when the Good is said to cause the being of the other Forms, it is not causing them to be the particular Forms they are, but rather it establishes them as having the sort of status they have. It causes them to be *simpliciter*, not to be this sort of thing as opposed to that. The content of the other Forms is not derived from the Form of the good. When the Form of the good causes the other forms to be, it simply causes them to exist. Οὐσία at *Republic* 509b8 means "reality," not "essence." The Form of the good causes the other Forms to exist and to have the status they have, not to have the essences which they have, or, more properly, *are*.

This simple existence of the Forms is bad news for both full-tilt Neoplatonists and Platonic predicationists. For Neoplatonist, because the content of Forms other than the Form of the good is not contained within the Form of the good; the other Forms are not emanations from it. For predicationists, because Plato has fully separated the sense of exist from other senses of "to be" and uses it to explain the status of Forms as distinct from the phenomena.

X. BEING AS BEING THERE

I defined existence for Plato as "being substantial," as "having the ability to shine forth," as "presence," as "being there in such a way as to be pointed at." These understandings admittedly purchase existence with dollars earned from perception and cognition. But backing them all is an understanding of existence as fundamentally a locative notion. This notion is not just Plato's but our own primitive, root understanding of existence as well. If we want to assert the existence of three sheep, we do not say, "Three sheep exist." Rather we say, "There are three sheep." And somewhere in this picture is a mind's finger to accompany the mind's eye, as though we were mentally pointing toward the sheep viewed against a 'field'.

In his summation of the nature of space, Plato comes very close to articulating place as his master metaphor for existence: "Space is that which we look upon as in a dream and say that anything that is must needs be in some place and occupy some room, and that what is not somewhere in earth or heaven is nothing" (*Timaeus* 52b3–5, Cornford).[16] On its face, the claim appears to be universal in scope: when *any* thing exists it does so because it is in some place. Some critics though have tried to limit the scope of this claim simply to the case at hand—that is, see it as applying just to the phenomena, which as images have to appear in something else on pain of being nothing at all (52c4–5): "It is not true for Timaeus that *everything* there is must be somewhere. For it is not the case that the forms are because they are somewhere in the receptacle. On the contrary, the forms do not even enter the receptacle (52a2–3)."[17] True enough, Forms are not in the receptacle, but that does not mean that metaphorically speaking they are not in space. The reference to being "somewhere in earth or heaven," given as a criterion for existence, looks to be just such a metaphor. The reference clearly extends beyond the case at hand. The phenomena exist not because they are "somewhere in earth or heaven," but because they are in the receptacle. And if we look through the dialogues, there are lots of places where the Forms are metaphorically treated as spatially extended. They exist "all in an order" (*Republic* VI, 500c), they exist on a plain (*Phaedrus* 248b), they exist in "heaven" or even in "the place beyond heaven" (*Phaedrus* 247c), they are "over there," "up there" (*Republic* VI 500d4, *Phaedrus* 250a2, 6). The receptacle of space provides a special form of "being there," the sort of "being there" that images have when their existence is made possible by a medium. But a Form has no need for a medium. It is there shining forth clearly on its own, and so may be said to exist completely.

X. GOD'S MIXING BOWL: SOME FINAL THOUGHTS ON DIVINE CRAFTSMANSHIP

The craftsman God of the *Timaeus* takes that which is without measure (ἀμέτρως, 53a8) and gives it measure (μέτρον, 68b6). This is how this

[16] At 71b–c, the unglamorous liver is viewed as a medium—explicitly compared to a mirror—for receiving and then broadcasting to consciousness images that constitute at least some of our dreams.

[17] Johansen 118.

maker makes. The principles of Demiurgic making are more abstractly formulated in the *Philebus* than in the *Timaeus*. Indeed the Demiurge himself at one point becomes a genderless abstraction, described by a series of neuter present participles—the making thing, the crafting thing (τὸ ποιοῦν, 27a5, τὸ δημιουγοῦν, 27b1). But the principles of demiurgic making in the *Philebus*—the introduction of measures and the elimination of excess and deficiency (23c–31a)—are the principles on which the Demiurge of the *Timaeus* operates. In the *Philebus*, the introduction of measure is typified by the creation of a precise number, a certain quantity (τὸ ποσόν, 24d5), say, the right temperature for a patient (26b6) or the summer (26b1). But these determinant quantities need not be raw integers (or otherwise single numbers, like 98.6). They may be fractions or ratios: "The equal and double . . . put an end to the conflict of opposites [the more and the less] with one another, making them well-proportioned and harmonious by the introduction of number" (*Philebus* 25d11–e2). It is these sorts of determinate quantities—ratios and proportions—that are the most prominent measures introduced by the Demiurge in the *Timaeus*—in the creation of the World Soul, other souls, the primary bodies, the World-Body, and heavenly bodies. But these measures' complex natures, complex, that is, when compared to a simple measure like 98.6° F., provide a portal into the text for those who (mostly non-reflectively) suppose that Plato thinks that to make (ποιεῖν), to put together (συνιστάναι), to (timber) frame (τεκταίνεσθαι), and to order (κοσμεῖν)—all processes repeatedly used to describe the Demiurge's fabrication of the world[18]—each involves either arranging determinate bits or forming unformed matter.[19]

In the *Timaeus*, the Demiurge uses no tools and only one utensil, a κρατῆρα (41d4), a mixing bowl, in which we are told, though only after the fact, that he had mixed up the components of the World-Soul and in which he goes on to mix up the ingredients of the souls of lesser beings. No critic believes that the mixing-bowl metaphor is to be taken literally. Still it is instructive for understanding the nature of demiurgic making in the *Timaeus*. The mixing bowl mentioned, κρατῆρα, is specifically a bowl for mixing water into wine in order to make the wine potable by the standards of decent people—not Scythians and other uncultured individuals who drink wine unmixed. Greek wines just by themselves were sickeningly sweet and too high of alcohol content to be drunk sanely in party portions. And so, just before being served, the wine was diluted

---

[18] Paronymous forms of all these verbs are used of the Demiurge, his activities, and derivatively of his creations in the *Timaeus*, e.g., 28b3, 28c3, 28c6, 29e1, just for starters.

[19] So, for example, Johansen translates ἀμέτρως at *Timaeus* 53a8 not as lacking "measure," but as lacking "order." Johansen 124.

with water, usually in a ratio of two or three parts of water to one part of wine. The establishment of this ratio of water to wine took place in this special utensil. The purpose of the mixing was not to create a third, new type of thing, an alloy, as the result of the two liquids transforming each other's nature.[20] The mixing was simply to get the water and wine in the right proportion to each other. As the water is poured into the wine there is always some fraction of the whole that the water is, just as the patient always has some degree of temperature as he gyrates between chills and fevers. The mixing process here simply aims at hitting upon the right fraction. That right fraction or proportion eliminates deficiency, say, a ratio of one to two, or one to one, and carefully avoids excess, say, a ratio of four to one or higher. The pouring and mixing does not create some new type of entity. The red stuff in the mixing bowl before the water was added was called "wine" (οἶνος) or "wine unmixed" (οἶνος ἄκρατος) or simply "unmixed" (ἄκρατος) and the red stuff that the drinkers actually drank was called simply "wine" (οἶνος—see *LSJ*, s.v.) . It is misleading then to think of the wine and water here as ingredients, in the way eggs, flour, sugar, baking powder, and chocolate are ingredients, ones which when mixed and fixed create a new type of entity, a cake, one that was not there before in the mixing bowl. So too the Demiurgic makings in the *Timaeus* should not be viewed as creating new types of entities, but simply getting right the proportions, fixing the numbers, of things that already exist.

The ratios and proportions that are the right numbers between things can be quite complex. They can be like the attunement or harmonizing of the stings of a guitar: each string is tuned individually by eliminating excess and deficiency in its tension, but in a way that integrates its right proportion, determinate number, to the modulated mores-and-lesses, the right proportions, of the other strings. The overall complex of proportions, proportions between proportions, could be called an order. And Plato uses the word τάξις (order) of the demiurgically created universe in both the *Timaeus* (30a5) and the *Philebus* (26b10). But there is order and there is order. And what is intended in these passages is an arrangement that consists of complex and compounded measures, not an arrangement either in the way a shape is imposed on clay or in the way determinate

[20] Indeed for some Platonic mixings, the original components can be retrieved from the mix. If you are going to plant a lawn, you want your seed to be a mix, a complex ratio, of several types of grass—Kentucky Blue, fescue, rye, maybe some Bermuda. Now, the seeds of all these grass types have different sizes and weights, and so could be separated back out from the mix by sieves or winnowing devices. Their mixed state just *is* their ratios. It is not a new type of grass, indeed, not a new type of anything. For Plato, even alloys can be undone, unmixed (*Statesman* 303d–e)

bits get arranged into a kind of thing that was not there before, as when Lego® blocks get arranged by a child into a castle. Instead Plato describes his complex measures, arrangements, or orderings as numbers set in relation to numbers, measures to measures (πᾶν ὅτιπερ ἂν πρὸς ἀριθμὸν ἀριθμὸς ᾖ μέτρον ᾖ πρὸς μέτρον, *Philebus* 25a8–b1). Plato's term of choice for such complex measures is συμμετρία (*Philebus* 25e1, 26a8). It would be wrong to think of συμμετρία here as meaning symmetry, balance, or commensurability—its etymological and common, dictionary senses. The term, though, does help move Plato's understanding of 'order' beyond the root understanding of τάξις—the disposition of an army—with its blunt suggestion that we are to think of order as an arrangement of determinate bits—hoplites. "Some assembly required."

Now we are in a position to better understand Plato's grand summation of the intersection of the works of reason and the effects of necessity. At *Timaeus* 69b–c, he writes:

> As we said at the outset, these things were in disorder (ἀτάκτως ἔχοντα) and god introduced into them all every kind of measure (συμμετρία) in every respect in which it was possible for each one to be in proportion and proper measure (ἀνάλογα καὶ σύμμετρα) both with iteself and with all the rest. For at first they were without any of these (proportion and measure), save by mere chance, nor was there anything deserving to be called by the names we now use—fire, water, and the rest; but all these he first set in order (διεκόσμησεν) and then put them together (συνεστήσατο) as this universe. (69b2–c1, Cornford adapted)

The ordering activity of Plato's god is neither Aristotelian nor atomistic. He neither imposes form on matter nor marshals atomic bits. He makes by reference to standards and measures and by introducing standards and measures into that which is neither intelligible nor good without them. Our thought styles are so thoroughly encased within Aristotelian and atomistic frameworks that is it hard for us to see that Plato provides us a different and interesting way of thinking about our world.

Plato thought that the best makings were emblemized by the builder, the carpenter, the timber framer. God is the timber-framing guy (ὁ τεκταινόμενος, *Timaeus* 28c6). Carpentry is the best of crafts because it "uses so many standards, measures, and instruments," and so makes things well, makes them fine, precise, and intelligible (*Philebus* 56b4–6, c4–6). Plato's gender and class kept him out of the kitchen and sewing room, or he might have chosen other crafts for his models of making. I first learned the Platonic way of viewing the universe—seeing it as consisting of measures, the measurable, and measurements taken from or imposed upon the measurable by the measures—not from reading the *Philebus* or hanging around job sites, but from pulling on my mother's apron strings and asking her "what's that for?" She raised three boys by teaching thousands of girls to cook and sew. Under the press of necessity, in a time when women did not typically work outside the home, she eked out a living in a lower-class junior highschool giving instruction in what then was pleonastically called "home economics" and now not called anything. But what a wonderful world it was—so many measures and standards and templates. There were dress patterns and french curves, measuring spoons and measuring cups, grains, drops, gills, pints, and pecks, candy thermometers, meat thermometers, oven thermometers, thermometer thermometers, stove timers, egg timers, hem-gages, needle-gages, pressure gages, tension sets and set squares, rulers, measuring tapes, yard sticks, scales, and a salinometer: "For roe, the reading should be 28.3." *The Joy of Cooking* wouldn't lie. How could I not become a Platonist. Thanks, Mom.

BIBLIOGRAPHY OF WORKS CITED

Ackrill, J. L. Review of Hackforth (2) *Mind* 62 (1953) 277–79.

Algra, K. *Concepts of Space in Greek Thought*. Leiden: Brill, 1995.

Allen, R. E., ed. (1) *Studies in Plato's Metaphysics.* London: Routledge, 1965. (Reprint with a new Introduction forthcoming in 2006 by Parmenides Publishing.)

———. (2) *Plato's 'Euthyphro' and the Earlier Theory of Forms.* London: Routledge, 1970.

———. (3) "Participation and Predication in Plato's Middle Dialogues" in Allen (1) 43–60.

———. (4) "The Argument from Opposites in *Republic* V" in Anton 165–75.

Ambrose, A. *Essays in Analysis*. London: Allen & Unwin, 1966.

Anton, J. P. and G. L. Kustas, eds. *Essays in Ancient Greek Philosophy.* Albany, N. Y.: State Univ. of New York Press, 1971.

Annas, J. (1) *Aristotle's Metaphysics, Books M and N.* Oxford: Clarendon, 1976.

———. (2) *An Introduction to Plato's Republic.* Oxford Clarendon, 1981.

———. (3) "Aristotle on Inefficient Causes" *Philosophical Quarterly* 32 (1982) 311–26.

———. (4) *Platonic Ethics, Old and New.* Ithaca.: Cornell Univ. Press, 1999.

Archer-Hind, R. D. *The Timaeus of Plato.* London: Macmillan, 1888.

Ayer, A. J. The *Problem of Knowledge*. New York: St. Martins Press, 1955.

Bambrough, R. "Universals and Family Resemblance," in M. J. Loux, ed. *Universals and Particulars: Readings in Ontology.* Garden City, N.Y.: Doubleday, 1970.

Bäumker, C. *Das Problem der Materie in der griechischen Philosophie*. Unaltered rpt. of 1890 edition. Frankfurt: Minerva, 1963.

Bluemel, C. *Greek Sculptors at Work.* 2nd English edition. Glasgow: Phaidon, 1969.

Brague, R. *Du temps chez Platon et Aristote.* Paris: Presses Universitaires de France, 1982.

Brandwood, L. (1) *The Chronology of Plato's Dialogues.* Cambridge: Cambridge Univ. Press, 1990.

———. (2) "Stylometry and Chronology" in R. Kraut, ed. *The Cambridge Companion to Plato.* Cambridge: Cambridge Univ. Press, 1992.

Broadie, S. (1) "Theodicy and Pseudo-History in the *Timaeus*" *Oxford Studies in Ancient Philosophy* 21 (2001) 1–28.

———. (2) "The Contents of the Receptacle" *Modern Schoolman* 80 (2003) 171–89.

Brown, L. "The Verb 'to be' in Greek Philosophy: Some Remarks" in S. Everson, ed. *Language: Companions to Ancient Thought* 3. Cambridge: Cambridge Univ. Press, 1994.

Brumbaugh, R. S. "Plato and the History of Science" *Studium Generale* 9 (1961) 520–27.

Burnyeat, M. F. "Platonism and Mathematics: A Prelude to Discussion" in A. Graeser, ed. *Mathematics and Metaphysics in Aristotle.* Bern: Verlag Paul Haupt, 1987.
Cherniss, H. F. (1) *Aristotle's Criticism of Plato and the Academy* I. Baltimore: Johns Hopkins Press, 1944.
———. (2) *The Riddle of the Early Academy.* Berkeley: Univ. of California Press, 1945.
———. (3) *Selected Papers.* Leiden: Brill, 1977.
———. (4) "The Philosophical Economy of the Theory of Ideas" in Cherniss (3) 121–32.
———. (5) "Plato as Mathematician" in Cherniss (3) 222–52.
———. (6) "The Sources of Evil According to Plato" in Cherniss (3) 253–60.
———. (7) "The Relation of the *Timaeus* to Plato's Later Dialogues" in Cherniss (3) 298–339.
———. (8)"*Timaeus* 38a8–b5" in Cherniss (3) 340–45.
———. (9)"A Much Misread Passage of the *Timaeus* (*Timaeus* 49c7–50b5)" in Cherniss (3) 346–63.
———. (10) Review of Festugière II in Cherniss (3) 455–67.
Cherry, R. S. "*Timaeus* 49c–50b" *Apeiron* 2 (1967) 1–11.
Claghorn, G. S. *Aristotle's Criticism of Plato's Timaeus.* The Hague: Nijhoff, 1954.
Clegg, J. S. "Plato's Vision of Chaos" *Classical Quarterly* 26 (1976) 52–61.
Cook Wilson, J. "On the Platonist Doctrine of the ἀσύμβλητοι ἀριθμοί" *Classical Review* 18 (1904) 247–60.
Cooper, J. ed. *Plato: Complete Works*. Indianapolis, Ind.: Hackett, 1997.
Cornford, F. M. (1) *Plato's Theory of Knowledge.* London: Routledge, 1935.
———. (2) *Plato's Cosmology.* London: Routledge, 1937.
———. (3) "The 'Polytheism' of Plato: An Apology" *Mind* 47 (1938) 321–30.
Crombie, I. M. *An Examination of Plato's Doctrines* II. London: Routledge, 1963.
Denniston, J. D. *The Greek Particles.* Oxford: Clarendon, 1934.
Dodds, E. R. (1) *The Greeks and the Irrational.* Berkeley: Univ. of California Press, 1951.
———. (2) "Plato and the Irrational" in *The Ancient Concept of Progress.* Oxford: Clarendon, 1973.
Easterling, H. J. "Causation in the *Timaeus* and *Laws* X" *Eranos* 65 (1967) 25–38.
Festugière, A.-J. *La Révélation d'Hermès Trismégiste* II and III. Paris: Gabalda, 1949, 1953.
Fine, G. (1) "Knowledge and Belief in *Republic* V" *Archiv für Geschichte der Philosophie* 60 (1978) 121–39.
———. (2) *Plato on Knowledge and Forms: Selected Essays*. Oxford: Clarendon, 2003.
Geach, P. T. "The Third Man Again" in Allen (1) 265–77.
Gill, C. and M. M. McCabe. *Form and Argument in Late Plato*. Oxford: Clarendon, 1996.
Gill, M. C. "Matter and Flux in Plato's *Timaeus*" *Phronesis* 32 (1987) 34–53.
Gosling, J. (1) "*Dóxa* and *Dúnamis* in Plato's *Republic*" *Phronesis* 13 (1968) 119–30.
———. (2) *Plato*. London: Routledge, 1973.
———. (3) *Plato: Philebus.* Oxford: Clarendon Press, 1975.
Gulley, N. "The Interpretation of Plato, *Timaeus* 49d–e" *American Journal of Philology* 81 (1960) 53–64.

Guthrie, W. K. C. *A History of Greek Philosophy* V. Cambridge: Cambridge Univ. Press, 1978.

Hackforth, R. (1) "Plato's Theism" in Allen (1) 439–47.

———. (2) *Plato's Phaedrus.* Cambridge: Cambridge Univ. Press, 1952.

———. (3) "Plato's Cosmogony (*Timaeus* 27d ff.)" *Classical Quarterly* 9 (1959) 17–22.

———. (4) *Plato's Philebus.* Cambridge Univ. Press, 1972.

Herter, H. (1) "Bewegung der Materie bei Platon" *Rheinisches Museum für Philologie* 100 (1957) 327–47.

———. (2) "Gott und die Welt bei Platon (Eine Studie zum Mythos des Politikos)" *Bonner Jahrbücher* 158 (1958) 106–17.

Hintikka, J. "Time, Truth, and Knowledge in Ancient Greek Philosophy" *American Philosophical Quarterly* 4 (1967) 1–14.

Irwin, T. (1) "Recollection and Plato's Moral Theory" *Review of Metaphysics* 27 (1974) 752–72.

———. (2) "Plato's Heracliteanism" *Philosophical Quarterly* 27 (1977) 1–13.

———. (3) "The Theory of Forms" in G. Fine, ed, *Plato.* Oxford: Oxford Univ. Press, 2000.

Johansen, T. J. *Plato's Natural Philosophy: A Study of the Timaeus-Critias.* Cambridge: Cambridge Univ. Press, 2004.

Kahn, C. H. (1) "Why Existence does not Emerge as a Distinct Concept in Greek Philosophy" *Archiv für Geschichte der Philosophie* 58 (1976) 323–34.

———. (2) "Some Philosophical Uses of 'to be' in Plato" *Phronesis* 26 (1981) 105–34.

Ketchum, R. "Plato on Real Being" *American Philosophical Quarterly* 17 (1980) 213–20.

Keyt, D. (1) "The Mad Craftsman of the *Timaeus*" *Philosophical Review* 80 (1971) 230–35.

———. (2) "Plato's Paradox that the Immutable is Unknowable" *Philosophical Quarterly* 19 (1969) 1–14.

Kneale, W. C. "Time and Eternity in Theology" *Proceedings of the Aristotelian Society* 61 (1961) 87–108.

Kung, J. "Why the Receptacle is not a Mirror" *Archiv für Geschichte der Philosophie* 70 (1988) 167–78.

Lee, E. N. (1) "On the Metaphysics of the Image in Plato's *Timaeus*" *Monist* 50 (1966) 341–68.

———. (2) "On Plato's *Timaeus,* 49d4–e7" *American Journal of Philology* 88 (1967) 1–28.

———. (3) "On the 'Gold-Example' in Plato's *Timaeus* (50a5–b5)" in Anton 219–35.

———. (4) "Reason and Rotation: Circular Movement as the Model of Mind (*Nous*) in Later Plato" in W. H. Werkmeister, ed., *Facets of Plato's Philosophy* (*Phronesis* Supplementary Volume II, 1976) 70–102.

Letwin, O. "Interpreting the *Philebus*" *Phronesis* 26 (1981) 187–206.

Lombard, L. "Relational Change and Relational Changes" *Philosophical Studies* 34 (1978) 63–79.

McCabe, M. M. *Plato's Individuals.* Princeton: Princeton Univ. Press, 1994.

Meldrum, M. "Plato and the 'ΑΡΧΗ ΚΑΚΩΝ" *Journal of Hellenic Studies* 70 (1950) 65–74.

Menn, S. *Plato on God as Nous*. Carbondale, Ill: Southern Illinois Univ. Press, 1995.
Miller, D. *The Third Kind in Plato's Timaeus*. Gottingen: Vandenhoeck & Ruprecht, 2003, *Hypomnemata* volume 145.
Mills, K. W. "Some Aspects of Plato's Theory of Forms, *Timaeus* 49c ff." *Phronesis* 13 (1968) 145–70.
Mohr, R. D. (1) "*Statesman* 284c–d: An 'Argument from the Sciences'" *Phronesis* 22 (1977) 232–34.
———. (2) "*Philebus* 55c–62a and Revisionism" in F. J. Pelletier and J. King-Farlow, eds., *New Essays on Plato* (*Canadian Journal of Philosophy* Supplementary Volume IX, 1983) 165–70.
Moravcsik, J. M. E. (1) "Recollecting Plato's Theory of Forms" in W. H. Werkmeister, ed., *Facets of Plato's Philosophy* (*Phronesis* Supplementary Volume II, 1976) 1–20.
———. (2) "Forms, Nature, and the Good in the *Philebus*" *Phronesis* 24 (1979) 81–101.
Morrow, G. R. "Necessity and Persuasion in Plato's *Timaeus*" in Allen (1) 421–37.
Mortley, R. J. "Primary Particles and Secondary Qualities in Plato's *Timaeus*" *Apeiron* 2 (1967) 15–17.
Mourelatos, A. P. D. "'Nothing' as 'Non-being': Some Literary Contexts that Bear on Plato" in J. P. Anton and A. Preus, eds., *Essays in Ancient Greek Philosophy* II. Albany, N.Y.: SUNY Press, 1983.
Natorp, P. *Platons Ideenlehre: Eine Einführung in den Idealismus*. Leipzig: F. Meiner, 1903.
Nehamas, A. (1) "Plato on the Imperfection of the Sensible World" *American Philosophical Quarterly* 12 (1975) 105–17.
———. (2) and Paul Woodruff. *Plato: Phaedrus*. Indianapolis, Ind.: Hackett, 1995.
Osborne, C. "Space, Time, Shape and Direction: Creative Discourse in the *Timaeus*" in Gill and McCabe, 179–211.
Owen, G. E. L. (1) "The Place of the *Timaeus* in Plato's Dialogues" in Allen (1) 313–38.
———. (2) "Plato and Parmenides on the Timeless Present" *Monist* 50 (1966) 317–40.
———. (3) "Plato on the Undepictable" in E. N. Lee, A. P. D. Mourelatos, and R. M. Rorty, eds., *Exegesis and Argument: Studies in Greek Philosophy Presented to Gregory Vlastos* (*Phronesis* Supplementary Volume I, 1973) 349–61.
Pappas, G. and M. Swain, eds. *Essays on Knowledge and Justification*. Ithaca: Cornell Univ. Press, 1978.
Parry, R. "The Unique World of the *Timaeus*" *Journal of the History of Philosophy* 17 (1979) 1–10.
Patterson, R. (1) "The Unique Worlds of the *Timaeus*" *Phoenix* 35 (1981) 105–19.
———. (2) *Image and Reality in Plato's Metaphysics*. Indianapolis, Ind.: Hackett 1985.
———. (3) "On the Eternity of Platonic Forms" *Archiv für Geschichte der Philosophie* 67 (1985) 27–46.
Press, G. A., ed. *Who Speaks for Plato? Studies in Platonic Anonymity*. Lanham, N.J.: Rowman & Littlefield, 2000.
Prior, W. J. *Unity and Development in Plato's Metaphysics*. LaSalle, Ill.: Open Court 1985.

Reed, N. H. (1) "Plato on Flux, Perception and Language" *Proceedings of the Cambridge Philological Society* 18 (1972) 65–77.
———. (2) "Plato, *Phaedrus* 245d–e" *Classical Review* 24 (1974) 5–6.
Reeve, C. D. C. *Philosopher-Kings*. Princeton: Princeton Univ. Press, 1988.
Robinson, T. M. (1) *Plato's Psychology.* Toronto: Univ. of Toronto Press, 1970.
———. (2) "The Argument for Immortality in Plato's *Phaedrus*" in Anton 345–53.
———. (3) "The Argument of *Timaeus* 27d ff." *Phronesis* 24 (1979) 105–109.
Rosen, S. "The Myth of the Reversed Cosmos" *Review of Metaphysics* 33 (1979) 59–85.
Ryle, G. "Plato's Parmenides" in R. E. Allen (1) 97–147
Santas, G. "The Form of the Good in Plato's *Republic*" *Philosophical Inquiry* 2 (1980) 374–403.
Schulz, D. J. *Das Problem der Materie in Platons Timaios.* Bonn: Bouvier, 1966.
Schwyzer, H. "Essence Without Universals" *Canadian Journal of Philosophy* 4 (1974) 69–78.
Sedley, D. "The Ideal of Godlikeness" in G. Fine, ed. *Plato*. Oxford: Oxford Univ. Press, 2000.
Shiner, R. A. *Knowledge and Reality in Plato's Philebus*. Assen: Van Gorcum, 1974.
Shorey, P. *The Unity of Plato's Thought.* Chicago: Univ. of Chicago Press. 1903.
Silverman, A. (1) "Timaean Particulars" *Classical Quarterly* 42 (1992) 87–113.
———. (2) *The Dialectic of Essence: A Study of Plato's Metaphysics.* Princeton: Princeton Univ. Press, 2002.
Skemp, J. B. (1) *The Theory of Motion in Plato's Later Dialogues.* 2nd and enlarged ed. Amsterdam: Adolf M. Hakkert, 1967.
———. (2) *Plato's Statesman.* London: Routledge, 1952.
Smith, N. "The Objects of Διάνοια in Plato's Divided Line" *Apeiron* 15 (1981) 129–37.
Stewart, J. A. *Plato's Doctrine of Ideas*. Oxford: Clarendon, 1909.
Strang, C. "Plato and the Third Man" in Vlastos (1) 184–200.
Strange, S. "The Double Explanation in the Timaeus" *Ancient Philosophy* 5 (1985) 25–39.
Tarán, L. "The Creation Myth in Plato's *Timaeus*" in Anton 372–407.
Taylor, A. E. (1) *A Commentary on Plato's Timaeus.* Oxford: Clarendon, 1928.
———. (2) "The 'Polytheism' of Plato: An Apologia" *Mind* 47 (1938) 180–99.
Verdenius, W. J. "Notes on Plato's *Phaedrus*" *Mnemosyne* 8 (1955) 265–89.
Vlastos, G., ed. (1) *Plato* I. Garden City, N.Y.: Anchor, 1971.
———. (2) "The Disorderly Motion in the *Timaeus*" in Allen (1) 379–99.
———. (3) "Creation in the *Timaeus:* Is it a Fiction?" in Allen (1) 401–19.
———. (4) *Platonic Studies.* Princeton: Princeton Univ. Press, 1973.
———. (5) "A Metaphysical Paradox" in Vlastos (4) 42–57.
———. (6) "Degrees of Reality in Plato" in Vlastos (4) 58–75.
———. (7) "Plato's Supposed Theory of Irregular Atomic Figures" in Vlastos (4) 366–73.
———. (8) *Plato's Universe.* Seattle: Univ. of Washington Press, 1975. (Reprint with a new Introduction by Luc Brisson published in 2005 by Parmenides Publishing.)
von Leyden, W. "Time, Number, and Eternity in Plato and Aristotle" *Philosophical Quarterly* 14 (1964) 35–52.

Wedberg, A. *Plato's Philosophy of Mathematics*. Stockholm: Almquist, 1955.
Whittaker, J. "The 'Eternity' of the Platonic Forms" *Phronesis* 13 (1968) 131–44.
Wittgenstein, L. (1) *The Blue and Brown Books*. New York: Harper & Row, 1958.
———. (2) *Philosophical Investigations*. Oxford: Basil Blackwell, 1958.
Zeyl, D. J. (1) "Plato and Talk of a World in Flux: *Timaeus* 49a6–50b5" *Harvard Studies in Classical Philology* 79 (1975) 125–48.
———. (2) *Plato: Timaeus*. Indianapolis, Ind.: Hackett, 2000.

INDEX OF PLATONIC PASSAGES CITED

AUTHOR INDEX